Geo. B. Selden in his "Benzine Buggy."

The present day motor car.

THE GASOLINE AUTOMOBILE

PREPARED IN THE
EXTENSION DIVISION OF
THE UNIVERSITY OF WISCONSIN

BY

GEORGE W. HOBBS, B. S.

INSTRUCTOR IN MECHANICAL ENGINEERING IN THE
UNIVERSITY EXTENSION DIVISION, THE
UNIVERSITY OF WISCONSIN

AND

BEN G. ELLIOTT, M. E.

ASSOCIATE PROFESSOR OF MECHANICAL ENGINEERING
THE UNIVERSITY OF NEBRASKA

©2011 Periscope Film LLC
All Rights Reserved
ISBN #978-1-935700-53-1
www.PeriscopeFilm.com

PREFACE

The purpose of this book is admirably expressed in the following quotation taken from the Buick instruction book: "To derive the greatest amount of satisfaction and pleasure from the use of his car the driver should have a complete understanding of the mechanical principles underlying its operation. Merely knowing which pedal to press or which lever to pull is not enough. The really competent driver should understand what happens in the various parts of the car's mechanism when he presses the pedal or pulls the lever. He should know the cause as well as the result."

When we consider the complexity of modern automobiles from a mechanical standpoint, with the duties that are required of them, together with the fact that the great majority of them are operated by men with little or no experience in the handling of machinery, the automobile stands as one of the most remarkable machines that the ingenuity of man has ever produced. The operating expense of the automobile has already assumed a large place in the budget of the American people. Although it is so built that the owner may secure good service from his automobile with very little knowledge of its construction, still it is evident that an intimate acquaintance with its details should enable him to secure better service at less expense and at the same time to prolong the useful life of the car.

It is with the hope of increasing the pleasure of automobile ownership and reducing the trouble and expense of operation that this book is offered. It is planned primarily for use in the University Extension work in Wisconsin, for the instruction of those who drive, repair, sell, or otherwise have to do with motor cars. It is largely the outgrowth of a series of lectures on the subject which were given in twenty-three cities of Wisconsin during the past winter.

The thanks of the authors are especially due to Mr. M. E. Faber of the C. A. Shaler Co. for assistance in preparing the section dealing with tire troubles, to Prof. Earle B. Norris for much of the chapter on Engines and for editing the manuscript and reading the proof, and to the many manufacturers who have liberally assisted in the preparation of the work by supplying their cuts and other material.

G. W. H.

MADISON, WIS.,
Sept. 15, 1915.

CONTENTS

CHAPTER I

General Construction

Art.		Page
1.	The steam propelled car	1
2.	The electric car	1
3.	The gasoline car	2
4.	Types of cars	2
5.	The chassis	2
6.	The frame	6
7.	The springs	6
8.	The front axle	8
9.	The steering gear	10
10.	The rear axle	12
11.	The differential	13
12.	The power plant and transmission	14
13.	The torque arm	15
14.	Strut rods	16
15.	Brakes	16
16.	Wheels	18
17.	Tires	19
18.	Rims	20
19.	The speedometer drive	21
20.	Control systems	23

CHAPTER II

Engines

21.	What is an explosion?	25
22.	Cycles	25
23.	The four-stroke cycle	26
24.	The order of events in four-stroke engines	27
25.	The mechanism of four-stroke engines	28
26.	Valve timing and setting	29
27.	Valves	30
28.	Valve arrangements	33
29.	The Knight engine	34
30.	The rotary valve	34
31.	Two-stroke engines	35
32.	The flywheel	38
33.	Ignition	39
34.	Clearance and compression	39
35.	Piston displacement	39
36.	Cylinder cooling	40
37.	The muffler	40
38.	Horse power of engines	41

CONTENTS

CHAPTER III

Power-Plant Groups and Transmission Systems

Art.		Page
39.	Single- and multi-cylinder engines	43
40.	Power plant and transmission arrangements	44
41.	Modern automobile power plants	50
42.	Constructional features of four- and six-cylinder engines	56
43.	Eight- and twelve-cylinder power plants	60
44.	Clutches	64
45.	Change gear sets	66
46.	Planetary gearing	67
47.	Universal joints and drive shaft	69
48.	Final drive	70
49.	Types of live rear axles	71

CHAPTER IV

Fuels and Carburetting Systems

50.	Hydrocarbon oils	75
51.	Fractional distillation of petroleum	75
52.	Principles of vaporization	76
53.	Heating value of fuels	79
54.	Gasoline gas and air mixtures	79
55.	Principles of carburetor construction	79
56.	Schebler, model L carburetor	82
57.	Schébler, model R	84
58.	The Holley model H carburetor	86
59.	Holley model G	87
60.	Stewart model 25	89
61.	Kingston model L	90
62.	Marvel carburetor	91
63.	Stromberg, model H	94
64.	Zenith model L	94
65.	Rayfield model G	95
66.	Carter model C	97
67.	General rules for carburetor adjustment	98
68.	Carburetor control methods	99
69.	The gravity feed system	99
70.	The pressure feed system	100
71.	The vacuum feed system	100
72.	Intake manifolds	102
73.	Care of gasoline	102

CHAPTER V

Lubrication and Cooling

74.	Friction and lubricants	103
75.	Cylinder oils	104
76.	Viscosity	104

CONTENTS

Art.		Page
77.	Flash point	104
78.	Fire test and cold test	104
79.	General notes on lubrication	104
80.	Splash system of engine lubrication	106
81.	Splash system with circulating pump	106
82.	Full forced feed system	111
83.	Mixing the oil with the gasoline	113
84.	Selection of a lubricant	113
85.	Directions for lubrication	114
86.	Cylinder cooling	117
87.	Water cooling systems	117
88.	Air cooling	122
89.	Cooling solutions for winter use	123

CHAPTER VI

Batteries and Battery Ignition

Art.		Page
90.	Fundamental electrical definitions	127
91.	Direct and alternating current	127
92.	Dry batteries	128
93.	Storage batteries	128
94.	Series and parallel connections	129
95.	Battery connections for ignition purposes	130
96.	Simple battery ignition system	130
97.	The three terminal coil	132
98.	Timers	135
99.	Spark plugs	135
100.	Master vibrators	136
101.	The high tension distributor system	137
102.	The Connecticut automatic ignition system	139
103.	The Atwater Kent system	141
104.	The Westinghouse ignition system	144
105.	The Delco system of ignition	147
106.	The Remy-Studebaker ignition system	149
107.	Spark advance and retard	151
108.	Automatic spark advance	151

CHAPTER VII

Magnetos and Magneto Ignition

Art.		Page
109.	Principles of magnetism	153
110.	Mechanical generation of current	155
111.	Low and high tension magnetos	156
112.	Armature and inductor types	156
113.	Remy model P magneto	157
114.	The Connecticut magneto	160
115.	Dual ignition systems	160
116.	Eisemann high tension dual ignition	161
117.	Eisemann automatic spark control	163
118.	The K-W high tension magneto	163

ART.		PAGE
119.	The Dixie magneto	166
120.	The Bosch high tension magneto	167
121.	The Bosch dual system	170
122.	Bosch two-independent system	173
123.	The Ford magneto and ignition system	174
124.	Magneto speeds	175
125.	Timing the magneto	176
126.	Battery vs. magneto ignition	177
127.	General suggestions on magnetos	177
128.	Common magneto ignition definitions	177

CHAPTER VIII

Starting and Lighting Systems

129.	Starting on the spark	179
130.	Mechanical starters	180
131.	Air starters	180
132.	Acetylene starters	180
133.	Electric starters	181
134.	Storage batteries	181
135.	Battery charging	185
136.	Wiring systems	187
137.	The Ward-Leonard system	187
138.	The Delco system	190
139.	Gray and Davis starting and lighting systems	193
140.	Wagner starting and lighting system	197
141.	The Westinghouse single-unit system	199
142.	Westinghouse two-unit system	200
143.	The U. S. L. electric starting and lighting system	204
144.	Jesco single-unit electric starter and lighter	205
145.	Care of starting and lighting apparatus	207
146.	Starting motor troubles	208
147.	Generator troubles	209
148.	Battery troubles	209
149.	Winter care of batteries	209
150.	"Don'ts" on starting equipment	210

CHAPTER IX

Automobile Troubles and Remedies

151.	Classification of troubles	213
152.	Power plant troubles	214
153.	Mechanical troubles in engine	216
154.	Carburetion troubles	221
155.	Ignition troubles	223
156.	Lubricating and cooling troubles	226
157.	Starting and lighting troubles	228
158.	Transmission troubles	228
159.	Chassis troubles	229

CONTENTS

CHAPTER X

Operation and Care

Art		Page
160.	Preparations for starting	231
161.	Cranking	231
162.	How to drive	232
163.	Use of the brakes	233
164.	Speeding	234
165.	Care in driving	234
166.	Driving in city traffic	235
167.	Skidding	236
168.	Knowing the car	237
169.	The spring overhauling	238
170.	Washing the car	240
171.	Care of tires	240
172.	Tire troubles	243
173.	Figuring speeds	247
174.	Interstate regulations	248
175.	Canadian regulations	249
176.	Touring helps–route books	250
177.	Cost records	250

Index 255

THE GASOLINE AUTOMOBILE

CHAPTER I

GENERAL CONSTRUCTION

Automobiles may be classified according to the type of power plant used, as steam, electric, and gasoline; or they may be divided into two classes according to use, as pleasure cars and commercial cars.

1. The Steam Propelled Car.—The steam engine has the advantage of flexibility. All operations such as starting, stopping, reversing, and acquiring changes of speed can be done directly by throttle control. By opening or closing the throttle, more or less steam is supplied to the engine, and the power is increased or decreased in proportion. When climbing a hill, all that is necessary to do is to give the engine more steam and consequently more power. The advantage of the steam engine in being able to start under load eliminates the clutch and also the transmission or change speed gears, the engine being connected directly to the rear axle.

The disadvantage of the steam engine is that it is necessary to fire up before starting, in order to generate enough steam to run the engine and propel the car. The steam machine requires large quantities of water to form the steam and that means frequent refilling of the water tank. They also require constant attention to the water and fuel pumps. The burning of the fuel under a boiler to generate the steam introduces an element of danger from fire and also makes the steam plant less efficient than the internal combustion engine.

2. The Electric Car.—The advantages of the electric car are similar to those of the steam car inasmuch as it is very flexible and can be controlled entirely by the controlling levers. By cutting out or in resistance, more or less current is supplied to the motor and the power of the motor is proportional to the flow of the current. The electric car is especially adapted to the use of women and children in cities. It is easy riding, clean, and very quiet.

The disadvantages are that it is not suitable for long drives, heavy roads, or hilly country. On one charge of the battery the average car will run from 50 to 100 miles, depending on the speed and condition of the roads. If the car is run at high speed, the battery will **not** drive the car as far as it will when running at moderate rate. This car is also limited to localities where there are ample facilities for charging the storage batteries.

3. The Gasoline Car.—The gasoline engine is much more economical than either the steam or electric, and after being once started has great flexibility. It is also better adapted for touring purposes than either of the others and does not require any more attention from the operator. The average car carries enough fuel to run it 200 to 400 miles without a stop and then it is necessary to fill the gasoline tank only, with an occasional quart or two of water for the radiator. With proper care, the engine will run as long as the gasoline supply and electrical system will hold out.

The disadvantages of the gasoline engine as compared with the steam engine or electric motor are, first, the gasoline engine is not self-starting; and, second, it lacks overload capacity. This means that some method of changing the speed ratio of the engine to the rear wheels is necessary in order to acquire extra power for climbing hills, for heavy roads, and also for reversing the car, as it is not possible to reverse the ordinary four-stroke automobile engine. The gasoline engine will not start under load, which necessitates the use of a clutch, so that the engine can be started and speeded up before any load is thrown on. Apparently there are a great many disadvantages to the gasoline engine but in reality they are very few, for with the proper handling of the spark and throttle controlling levers it is not necessary to keep continually changing gears. The speed change lever need not be used except for starting, stopping, hill climbing, and on bad roads.

4. Types of Cars.—In general, the parts of the pleasure and commercial cars are the same except that the pleasure cars are built much lighter than the commercial cars. In the pleasure car everything is planned for comfort and speed, while the commercial car is built for heavy loads and is generally intended to be driven at low speed.

The principal body types of pleasure cars are, *the limousine, the touring car, the coupé,* and *the roadster*, as shown in Fig. 1.

The commercial cars are built for light, medium, and heavy duty. A few of the commercial types are shown in Fig. 2.

The cycle car is a name commonly given to small cars which have less than 70 cu. in. piston displacement or a tread of less than 56 in.

5. The Chassis.—The principal parts of the gasoline automobile are the frame, springs, axles, wheels, power plant and auxiliaries, clutch, transmission system, controlling apparatus and body. The chassis, as shown in Fig. 3, includes all parts with the exception of the body and its accessories. The functions and types of these parts will be taken up separately.

GENERAL CONSTRUCTION

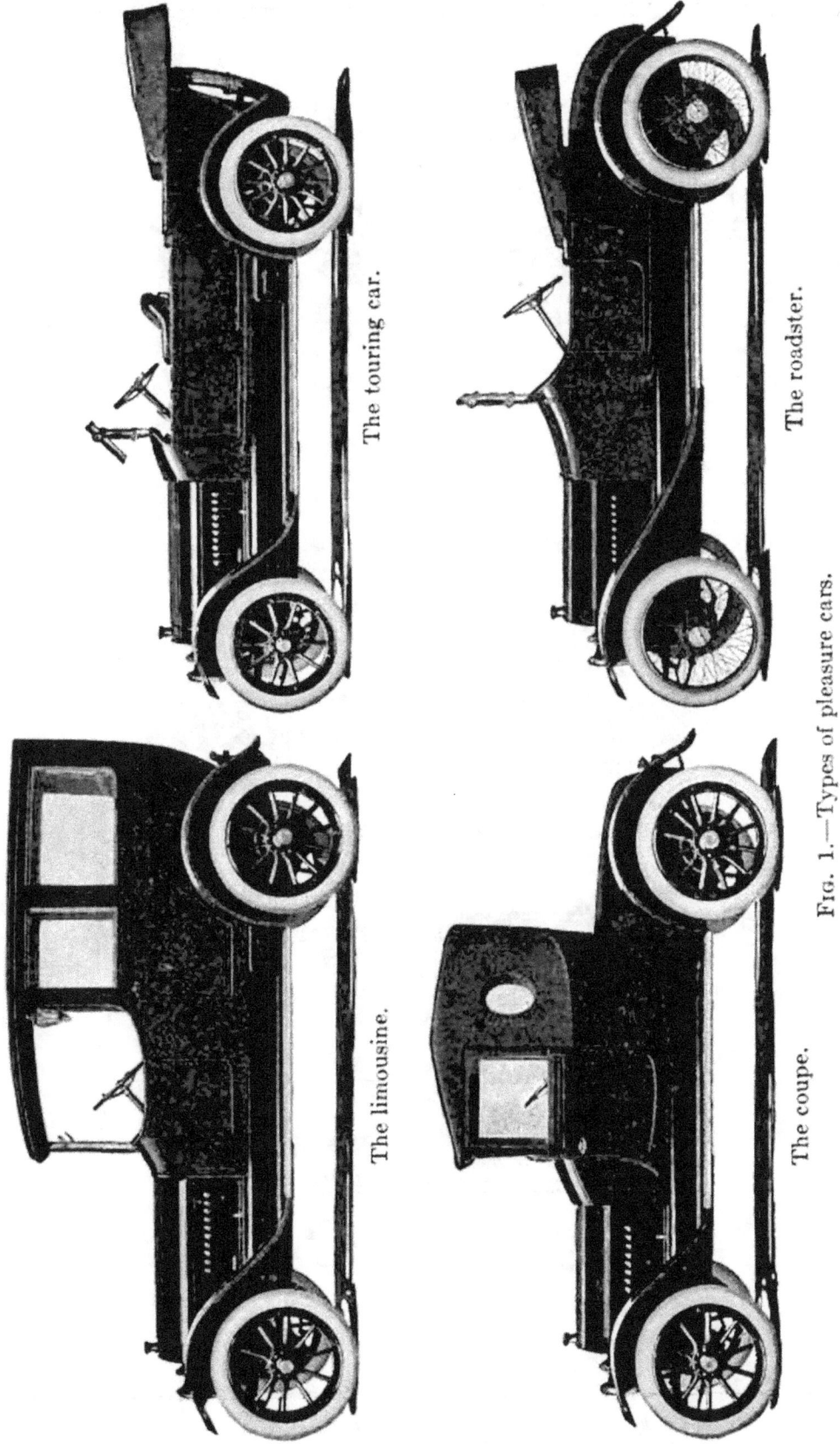

Fig. 1.—Types of pleasure cars.

4   THE GASOLINE AUTOMOBILE

FIG. 2.—Types of commercial cars.

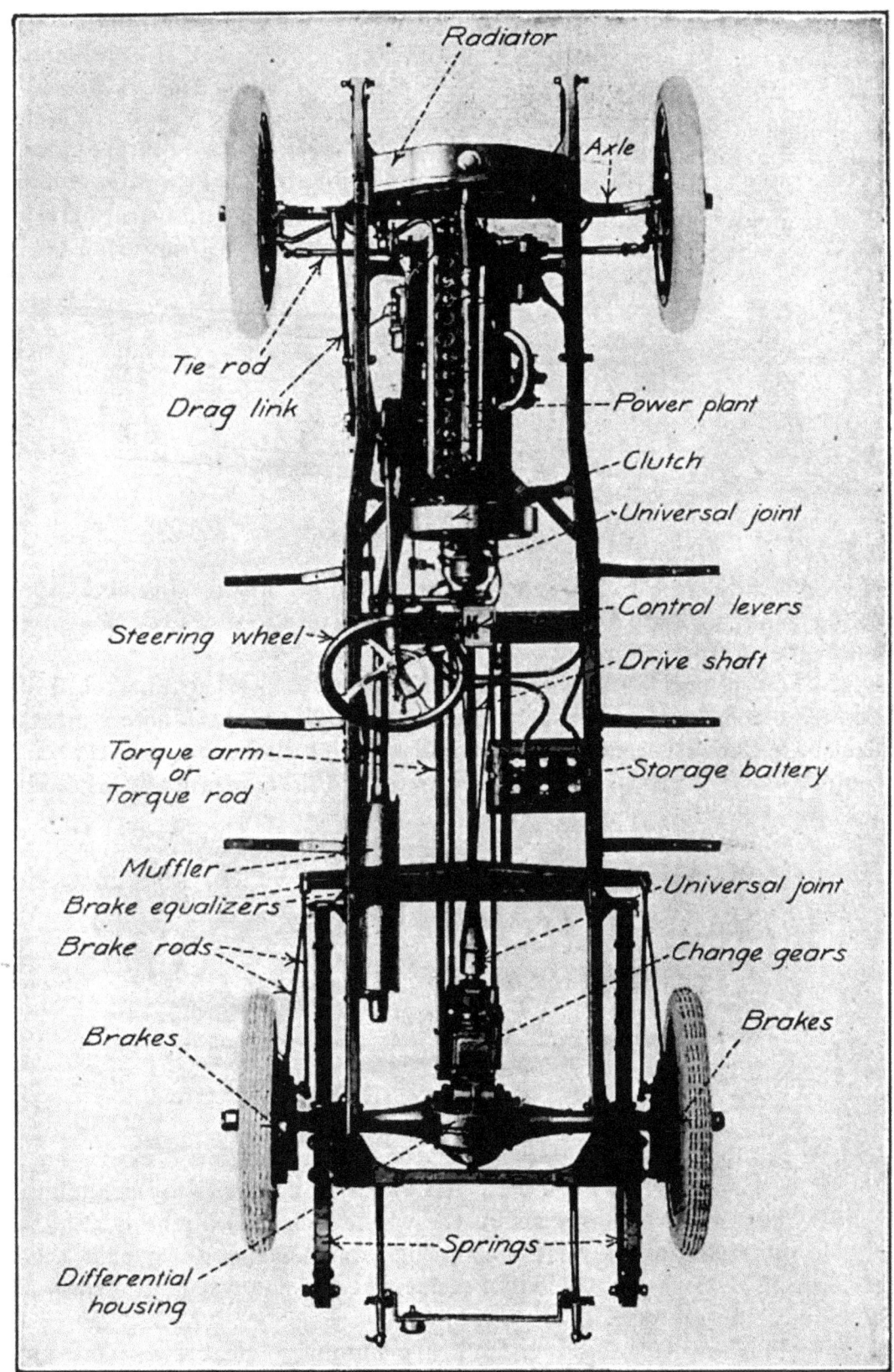

Fig. 3.—Chassis of the Studebaker "Six."

6. The Frame.—The automobile frame is a very important part of the car, due to the fact that it supports the power plant, transmission mechanism, body, etc. The frame is attached to the springs, which in turn are fastened to the axles. Frames are made either of wood or metal or a combination of the two. The metal frames are usually of channel-section steel. The wooden frames may be either of the solid timber type or of laminated strips glued together and sometimes reinforced by steel strips. This type is very strong and light and does not transmit so much

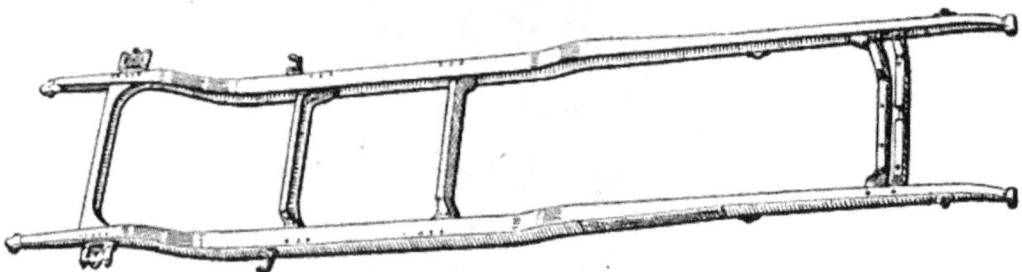

Fig. 4.—Channel steel frame.

of the vibration as the steel frame. Figure 4 shows a pressed steel channel-section frame. Figure 5 shows a frame made from second-growth ash and used on the Franklin car.

7. The Springs.—The frame of the automobile is supported by laimated leaf springs. Coil springs are used only in places where a great deal of strength is needed in a small space and where quick action is required. The springs under the frame of an automobile must be gradual

Fig. 5.—Franklin wood frame construction.

and easy in their action, and this is why the laminated leaf spring is used. The strength and resilience of the leaf spring can be varied by changing the number of leaves or by varying the width or length of the leaf. It also has an advantage over the coil spring in that if one leaf breaks the spring is still serviceable, while in a coil spring if a coil breaks the spring is no longer of any use.

The laminated spring is built up of a number of leaves varying in length, the longest leaf being on the concave side of the spring and the

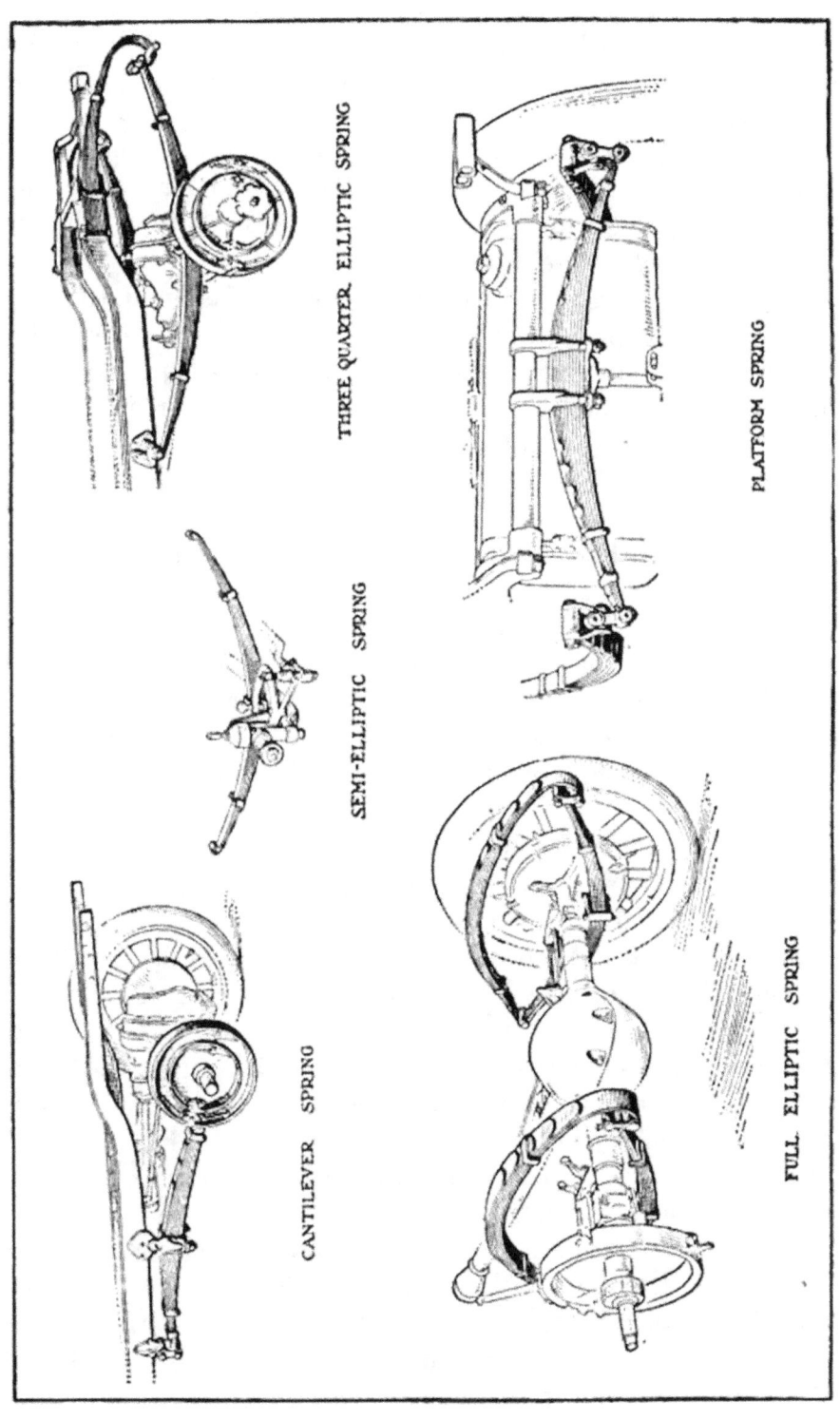

Fig. 6.—Types of Springs.

other leaves built on this one in the order of their length. The ends of the long leaf are bent around to form eyes so that they can be fastened to the frame by a clevis or other means.

The laminated leaf springs, as shown in Fig. 6, are built in the following forms: cantilever, semi-elliptic, three-quarter elliptic, full-elliptic, and platform springs.

The cantilever spring is fastened flexibly to the frame at one end and the center and carries the axle at the other end. There is another type of cantilever spring which has a single rigid fastening to the frame. This is also called a quarter-elliptic spring.

The semi-elliptic spring usually has its center fastened to the axle while the two ends support the frame. This type of spring is generally used to support the front end of the car, because this type has the least amount of side-sway. Since the front axle is used for steering purposes, a great amount of flexibility is not desired.

The three-quarter elliptic spring consists of a semi-elliptic member, to one end of which is attached a quarter-elliptic member. This type is supported in the middle of the semi-elliptic spring and is connected to the frame at one end of the semi-elliptic and the free end of the quarter-elliptic springs.

The full-elliptic spring consists of two semi-elliptic springs connected together at the end, supported at the middle of one semi-elliptic and carrying the load at the middle of the other. Either the three-quarter or the full-elliptic types have greater flexibility than the semi-elliptic type.

The platform spring consists of three semi-elliptic springs fastened together. Two of the members are parallel to the sides of the car and the third is inverted and is parallel to the cross members. The car frame is attached to the front end of the side members and to the middle of the cross member. The middle of the side members rests on the spring seats.

8. The Front Axle.—The front axle consists of the center, the knuckles, a steering arm, a third arm, a plain arm, and the tie rod. The centers are either I-beam, as shown in Fig. 7 or tubular as in Fig. 8, and they may be either straight or dropped center types. Square centers are sometimes used on heavy trucks.

The I-beam centers are made either of drop forgings or of cast steel and are heat-treated to do away with brittleness and give strength and toughness. The tubular centers and tie rods are made from the best high-grade seamless steel tubing and the yokes are either pinned or brazed on the ends of the tubes. In the I-beam centers the yokes form a part of the forging or casting. The I-beam construction is the strongest but is not quite so flexible as the tubular center.

GENERAL CONSTRUCTION

The front wheels are fastened on the spindle of the knuckle and run on cup-and-cone ball bearings or on roller bearings as shown in Fig. 7. The spindle is set so that the front wheels have a camber of about 2 in., that is, the tops of the wheels are about 2 in. farther apart than the

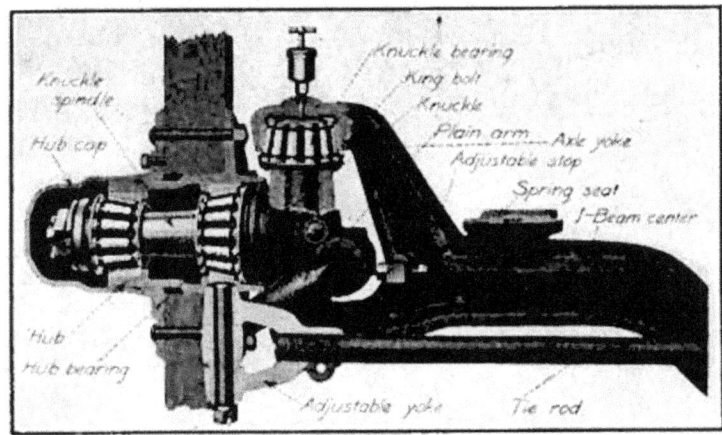

Fig. 7.—I-beam front axle construction.

bottoms of the wheels. This is to conform to the crown of the road and to bring the point of contact between the tire and the road in line with the king-bolt.

In order to make the car steer easier and have a tendency to run straight ahead, the front wheels should toe in from ⅜ to ½ in. This is done by adjusting the length of the tie rod.

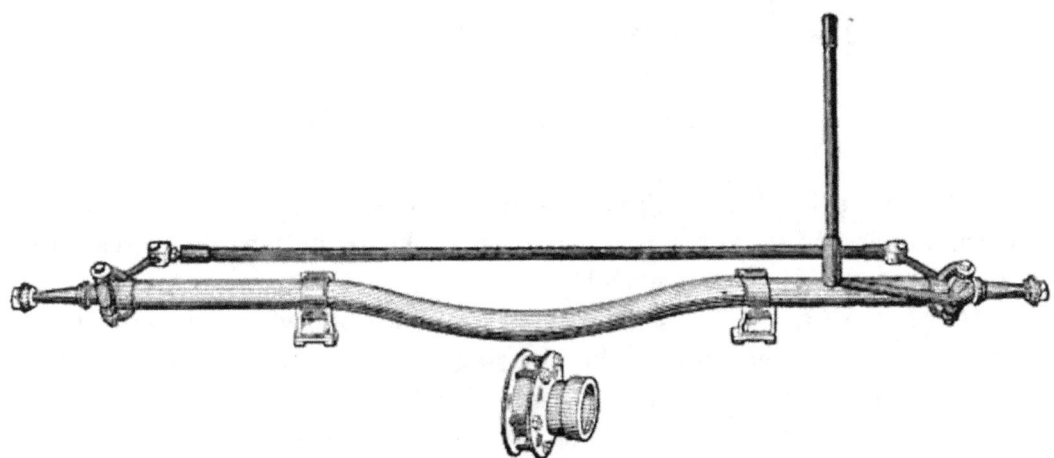

Fig. 8.—Tubular front axle.

The knuckles are fastened in the axle yokes by king-bolts and are free to swing about 35° either way from the center line of the axle. This is necessary in order to allow the wheels to follow a curve when turning. Between the top of the axle yoke and the knuckle there should be a ball or roller bearing or a renewable bronze washer to carry the load and yet allow the knuckle to turn easily.

The king-bolt should fit in a bronze bearing in order to insure easy movement and a small amount of wear. The steering and third arms, which are generally combined in a single forging, are keyed to one knuckle. The third arm is connected by the tie rod to the plain arm, which is keyed to the other knuckle. The general layout of the steering apparatus is shown in Fig. 9. The steering arm is connected by the drag link to the pitman arm or steering lever on the base of the steering gear.

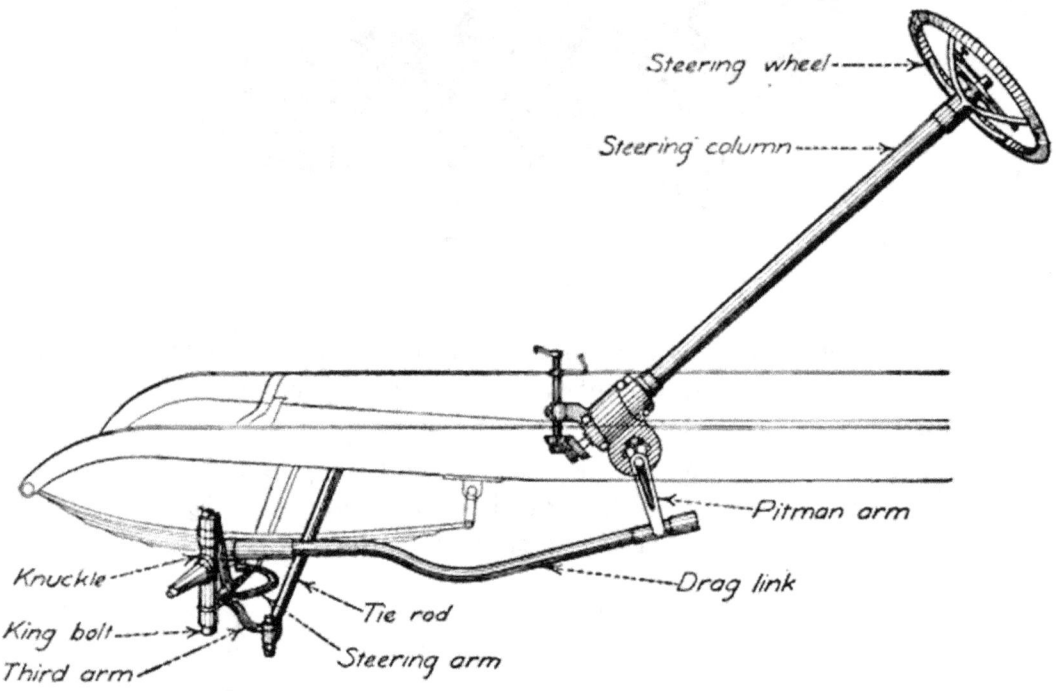

Fig. 9.—Arrangement of steering apparatus.

9. The Steering Gear.—The steering gear is the part of the mechanism that operates on the knuckles to turn the front wheels in response to movements of the hand wheel.

Figure 10 shows the essential parts of a double worm steering gear. Inside the steering column is the steering tube, the upper end of which is connected to the hand wheel while the lower end carries a double-threaded worm. The worm meshes with two half-nuts, one with a right-hand and the other a left-hand thread. Two rollers, which are attached to the yoke that operates the pitman arm or steering lever, bear against the lower ends of the half-nuts. The operation is as follows: Turning the hand wheel turns the tube and worm in the same direction, which causes one half-nut to rise and the other to descend. This pushes one roller down and lets the other rise. The yoke is given the same motion and transmits it to the pitman arm, which pushes or pulls on the drag link and thus turns the knuckle and wheels.

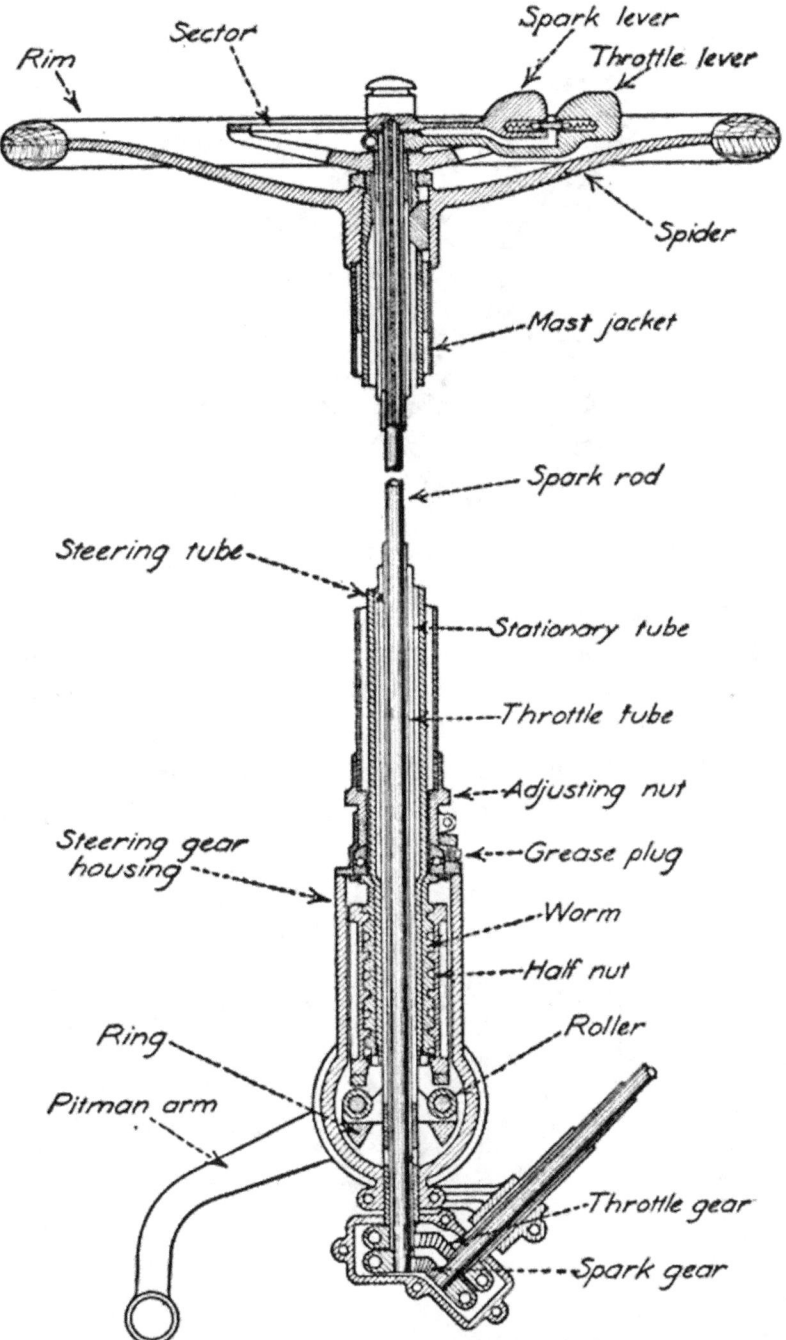

Fig. 10.—Double worm steering mechanism.

Figure 11 shows the worm-and-gear type. The worm is fastened to the steering tube and is turned with the hand wheel. The gear shaft carries the pitman arm, which connects to the knuckle steering arm by the drag link.

These steering gears are non-reversible, because while the action of the hand wheel is readily transmitted to the front wheels the jarring of the front wheels on rough roads can not be transmitted back to turn the hand wheel.

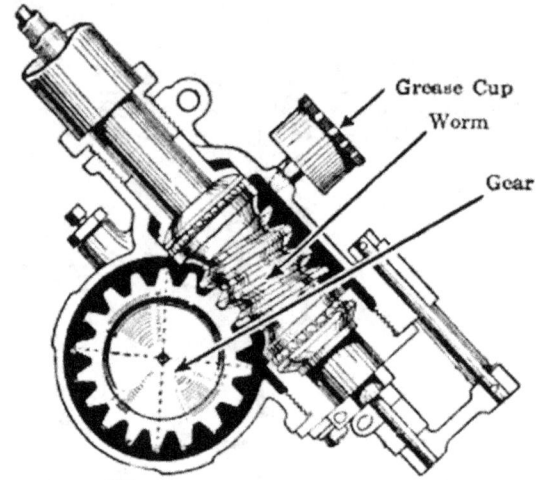

Fig. 11.—Worm-and-gear steering mechanism.

10. The Rear Axle.—The rear axle must carry this end of car and also provide means of giving power to the rear wheels to propel the car. This is done in two general ways, and the corresponding types of axles are called "dead" and "live" axles.

Figure 12 shows a truck chassis with a dead rear axle. It is somewhat similar in construction to the ordinary wagon axle, as it is made up of a

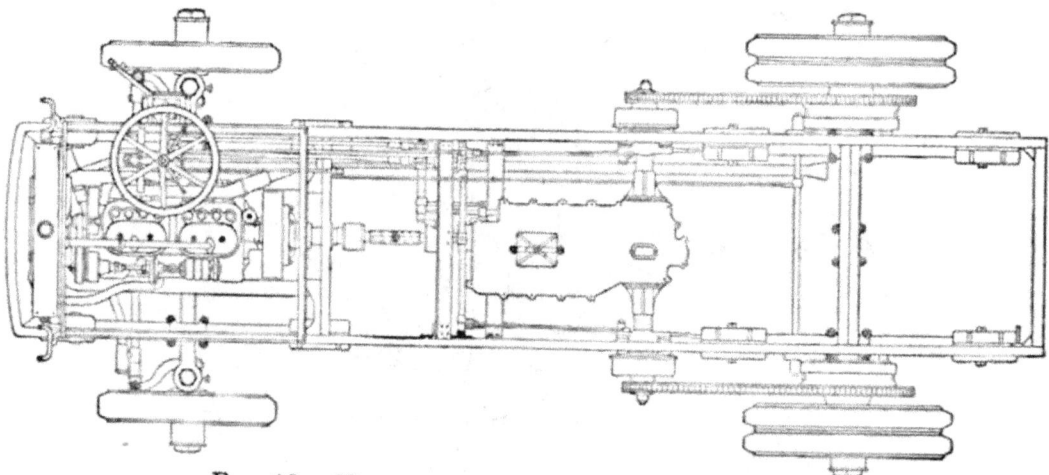

Fig. 12.—Heavy truck chassis with dead rear axle.

solid bar with spindles machined on the ends for the wheel bearings. The wheels have large sprockets on the inside which are driven by chains from other sprockets on the ends of a "jackshaft" near the middle of the car. This type of axle is used principally on heavy trucks where it is necessary to have a solid construction and provide for a large reduction in speed.

For pleasure cars, the live axle is generally used. The general arrangement of a car with a live axle was shown in Fig. 3. In Fig. 13 is shown in detail the construction of a typical live axle. In this type the axle turns and drives the rear wheels with it. The axle is surrounded by a stationary housing which supplies the bearings for the wheels and the axle and which also supports the car through the springs. The live axle receives its power near the center, usually through a set of bevel gears which give the desired speed reduction and also make the necessary right angle change in the power transmission.

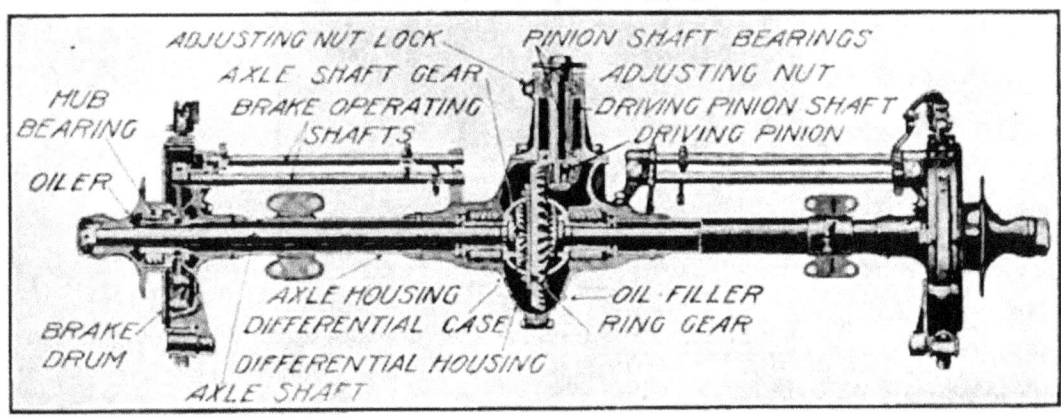

Fig. 13.—Live rear axle.

11. The Differential.—Some provision has to be made to drive the rear wheels positively in either direction and yet allow one wheel to run ahead of the other when turning a corner. This is done by dividing the live rear axle at the center and connecting the two halves by a differential gear, the details of which are shown in Fig. 14. Each half of the live axle (called the *main shaft* in Fig. 14) has a bevel gear on its inner end. These bevel gears face each other and are called the *differential gears*. They are connected by from two to four *differential pinions* spaced at equal distances around the circle. The power is applied at the centers of these differential pinions so that they act like the doubletrees or eveners on a team of horses, allowing one wheel to run ahead of another or to lag behind but still maintaining an even pull on the two differential gears. Referring to Fig. 14, the power from the engine is brought back to the driving pinion and this delivers it to the large gear called the *bevel ring*. This bevel ring is fastened to the *differential case*, which, therefore, receives the power from the bevel ring. The differential case turns the *spider* with it and, as this spider carries the differential pinions, these pinions are carried around with a force applied at their centers. On a straight road the differential case, the spider, the differential pinions and the differential gear all revolve as one mass and there is no internal action in the differential. The differential pinions pull equally on the two differential gears on each side of them and they all revolve together. In

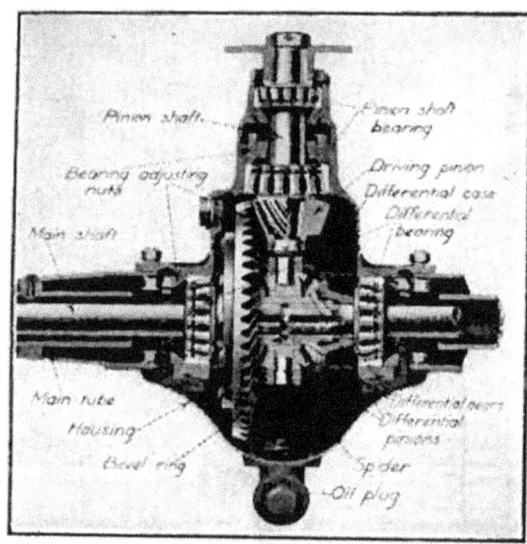

Fig. 14.—Differential gear.

turning a corner the outer wheel has farther to go and hence must run faster. This makes the one differential gear turn faster than the other. This causes the differential pinions to revolve on their axes, but they still continue to deliver power equally to the two wheels.

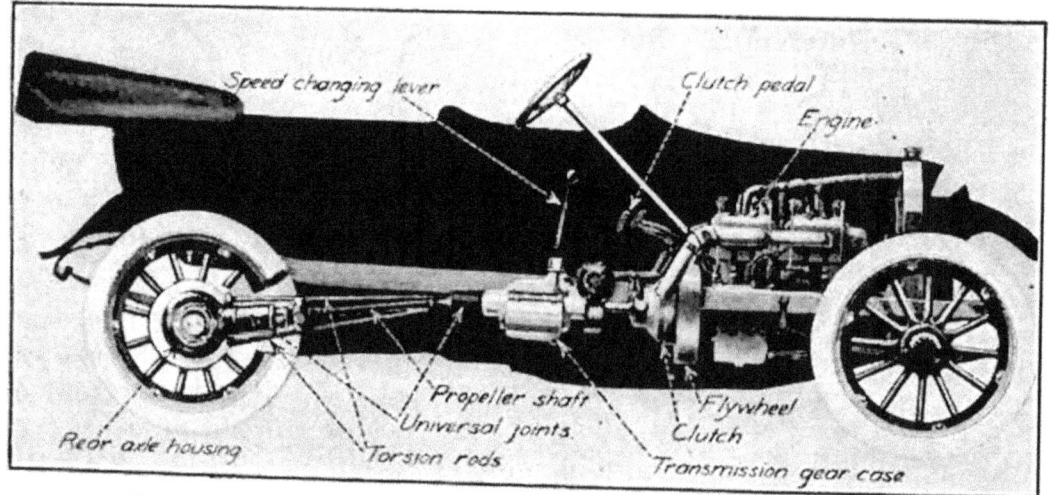

Fig. 15.—Arrangement of power plant and transmission system.

12. The Power Plant and Transmission.—Figure 15 shows a typical arrangement of the power plant and the power transmission system. The engine is generally placed in the front end of the car, both for accessibility and to balance the weight of the passengers in the rear part of the car. The engine is the most important part of the car. Its purpose is to transform the heat energy of gasoline into mechanical energy at the crank shaft for the purpose of driving the car. The power is delivered to the flywheel, from which the clutch takes it and passes it back to the transmission. In the transmission case is a system of gears for reducing the speed from the engine and increasing the turning force for starting purposes or for heavy driving, as in sand or on hills.

The power plant is mounted on the frame of the car, while the rear wheels which are to finally receive and use the power are flexibly connected to the frame by springs. We must, therefore, have a flexible arrangement for taking the power from the power plant to the rear axle. This is usually accomplished by means of a propeller shaft and one or two universal joints (see Fig. 15). A universal joint is merely a double-hinged shaft connection (see Fig. 16) permitting the lower end of the propeller shaft to swing at will with the rear axle and yet receive power from the engine.

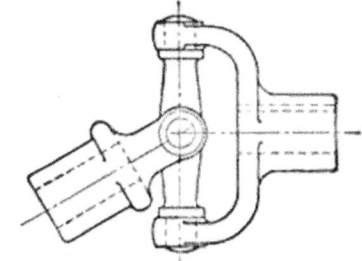

Fig. 16.—Universal joint.

In the car of Fig. 15 the engine and transmission are carried in the frame of the car and the first universal lies just back of the transmission. In the car of Fig. 3 the transmission with its change gears is placed just in front of the rear axle and is fastened solidly to the rear axle housing. This places both universal joints and the propeller shaft between the engine and the transmission.

In addition to the engine proper, the power plant contains a number of accessories necessary for the operation of the engine, such as the lubricating system, the ignition system, the carburetor, the cooling system, and the starting system. In the so-called unit power plant the clutch and change gears are contained in a single unit with the engine. All these accessories will be taken up in the later chapters.

In heavy trucks the system of power transmission is somewhat different from the pleasure car system just described. The power from the engine is carried through the clutch and back to the transmission located in the center of the chassis, as shown in Fig. 12. Here the power is turned at right angles in the rear part of the transmission and is given to a jackshaft lying across the car. The sprockets on the outer ends of this jackshaft drive the rear wheels through two chains. No universal joints are needed in the final drive, as the chains allow for the free motion of the rear axle.

13. The Torque Arm.—When the brakes are used in stopping a car, the brakes, being carried by the rear axle housing, tend to carry this housing around with the wheels. Likewise, the action of the propeller shaft and the bevel pinion in driving the rear axle (see Fig. 14) tend to turn the axle housing over backward with the same force that is exerted on the bevel ring. This twisting action or "torque" must be taken care of in some way. This can be done by torsion rods as in Fig. 15, or by a single bar called a torque arm or by a torsion tube around the propeller shaft, or it can be left entirely to the springs to take care of this action. If the torque is taken up by a housing around the propeller shaft as in Fig. 17, this tube is called the "third member" of the rear axle system and is securely bolted to the rear axle housing. This system does away with one universal joint, as only one at the front extremity of the propeller shaft is used.

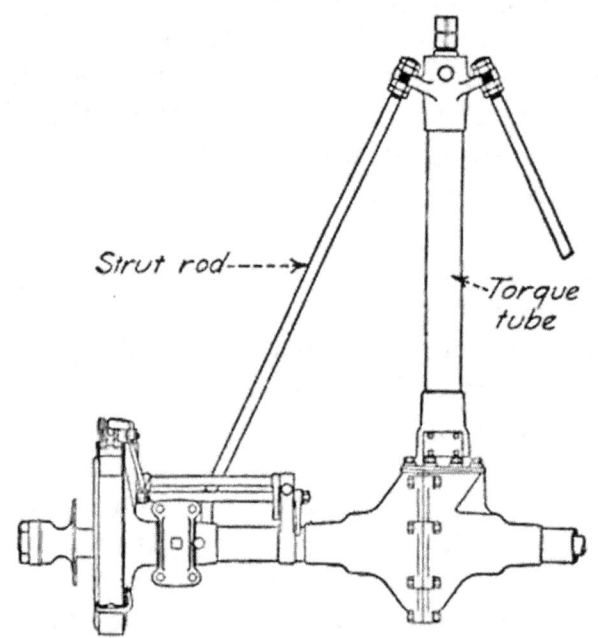

FIG. 17.—Rear axle with torque tube and strut rods.

14. Strut Rods.—In order to preserve the alignment of the wheels or to keep one wheel from getting ahead of the other, strut rods are fastened to the brake flanges or spring seats, and extend to the front end of the third member as in Fig. 17 or to some part of the frame.

15. Brakes.—Brakes which act on the rear wheels are either of the contracting or expanding band type or the expanding shoe type.

Figure 18 shows the general layout. This is known as a double internal type of brake. A steel brake drum is fastened securely to the wheel. Both bands expand and put pressure on the inside of the drum. The outside band, or the one next the wheel, is the emergency brake and is operated by a hand lever. The other, the service brake, is under the control of the driver through the medium of the foot pedal. The brake bands are carried by brake flanges near the ends of the rear axle housing. The two sets are entirely independent of each other. Another type of internal expanding band brake that uses two brake drums is shown in Fig. 19. The action is similar to the above. In this case the smaller

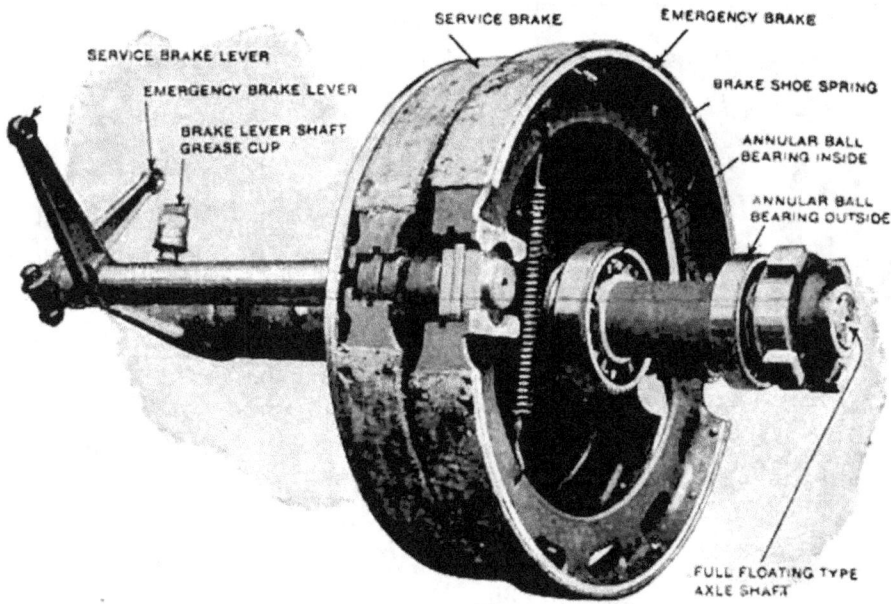

FIG. 18.—Double internal brake with single drum.

FIG. 19.—Double internal brake with two drums.

band is used for the emergency. Figure 20 shows a type of brake known as the internal-external brake. There are two bands working on the same drum. One set contracts around the outside of the drum and the other set expands against the inner circumference. The outer band constitutes the service or foot brake and the inner band the emergency brake.

Fig. 20.—Internal-external brake.

All bands, either contracting or expanding, are faced on the rubbing side with an asbestos preparation that is capable of standing a great amount of wear and is not easily burned out. Some types that use the expanding shoe have a cast-iron shoe that is pressed against the inside of the steel drum on the wheel.

A typical mechanism for operating the expanding shoes or drums is clearly shown in Fig. 18, where the emergency band is shown expanded while the service brake is in the running position.

16. Wheels.—Automobile wheels are classified as artillery wheels (with wooden spokes), wire wheels, and cast- or pressed-steel wheels, the latter being limited to heavy duty trucks.

Artillery Wheels.—The artillery wheel, shown in Fig. 21, is built of second-growth hickory. The spokes are fastened together at the

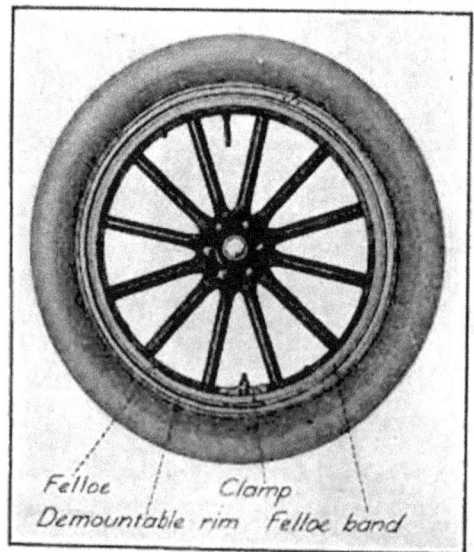

Fig. 21.—Artillery wheel.

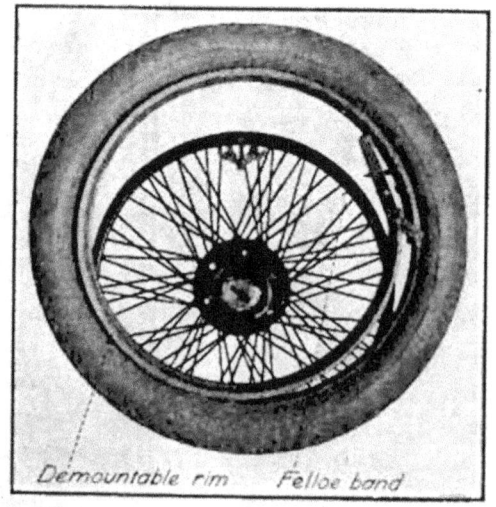

Fig. 22.—Wire wheel.

hub of the wheel by a series of interlocking mortise-and-tenon joints and the outer ends are turned down to fit in holes in the wooden felloe band.

The hub casting, which serves to hold the inner end of the spokes, also acts as the bearing housing for the hub bearings, on which the wheel revolves.

Wire Wheels.—The wire wheel is shown in Fig. 22. On account of the scarcity of second-growth hickory, which is the only acceptable material for artillery wheels, some companies are building wire wheels which are modifications of the bicycle wheel. Wire spokes are interlaced between the hub and rim in such a manner that the wheel is held rigid and withstands both the direct loads and side strains.

In the artillery wheel, the load is carried by the spokes on the under side. In the wire wheel, the load is carried by the spokes above the hub.

The advantages claimed by the wire wheel manufacturers are that the wheel is reduced in weight about 30 per cent.; is more resilient, which makes an easier riding car; will stand greater radial strain; and is fully as strong as the artillery wheel.

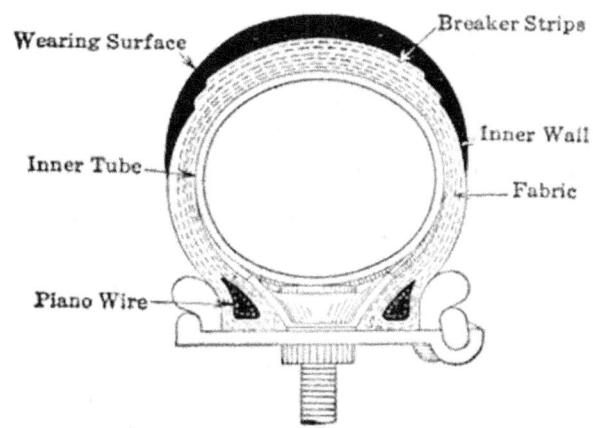

Fig. 23.—Section of pneumatic tire.

17. Tires.—The tires used on pleasure cars are usually of the pneumatic rubber type. Some are being filled with a spongy substance that makes them more of a cushion form and some have bridges of para rubber instead of an air cushion. The lighter commercial cars use solid rubber tires, the heavier trucks use steel tires, while some are using wooden blocks. The wooden blocks and steel tires can be used only on the very low-speed trucks on account of there being no resilience in tires of these types.

The pneumatic tire serves as a good shock absorber and eliminates a large portion of the road vibrations and jars before they reach the mechanism of the car.

The general construction of the tire is shown in Fig. 23. Several layers of heavy canvas (friction fabric) are wound around two circular wire cables (beads) in the shape of a tire. This forms the foundation, which is filled with rubber gum to form the carcass of the tire. Around the carcass the cushion is built, which is an extra thickness of compounded rubber held in place by a double layer of canvas. This is called the breaker strip. Outside of this comes the tread. The tread is the part that comes into contact with the road and takes the wear. This whole structure is then vulcanized to make a solid unit.

The inner tube, which is merely a rubber bag with a check valve to hold the air, is inserted in the casing and the casing is fitted on the

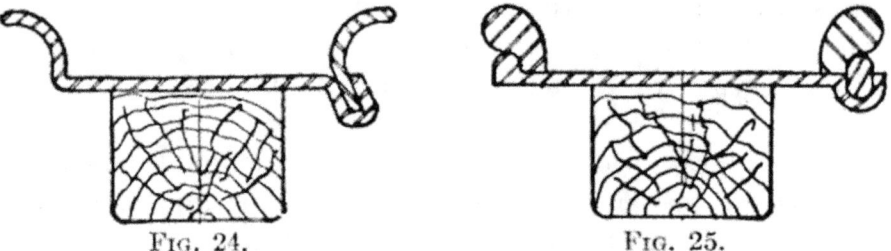

FIG. 24. FIG. 25.

FIGS. 24 AND 25.—Types of detachable rims.

rim in such a way that when the pressure is applied the bead grips the rim, and the flanges on the rim prevent the tire from sliding off sideways.

18. Rims.—Rims may be classified as clincher, detachable, and demountable, or a combination of two of these. The cuts shown in Figs. 24, 25, and 26 show sections of the Goodyear rims. Figure 24 illustrates the detachable rim of two parts. The side ring can be easily removed from the groove by a screw-driver. The higher the inflation pressure in the tire the harder the side ring hugs the groove. This rim is used to a great extent on electric pleasure cars.

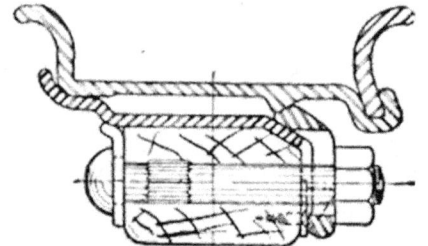

FIG. 26.—Demountable-detachable rim.

Figure 25 shows a heavier type of detachable rim, quite general on gasoline pleasure cars.

Figure 26 shows a rim which has both the demountable and detachable features combined. With demountable rims, an extra rim with tire fully inflated may be carried. In case of a blow-out, the damaged tire and its rim may be quickly removed and the spare rim and tire put on. This saves considerable time in cases of tire trouble.

Figures 27 and 28 show the rim made by the General Rim Co. This is a demountable rim and is locked on the rim at a single point. To remove the rim from the wheel the toggle nut is turned to its lowest position on the end of the clamping bolt, as shown in Fig. 28. This draws the clamping ring into the groove and the rim is released and ready for removal. To replace the rim merely reverse this operation.

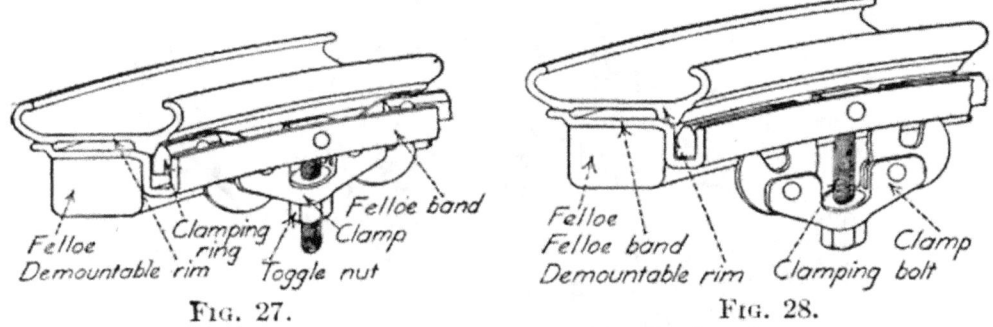

FIG. 27. FIG. 28.
FIGS. 27 AND 28.—Demountable clincher rim.

Figure 29 shows sections of the clincher rim as used on the Ford car, and also shows the method of removing the tire from the rim.

19. The Speedometer Drive.—Some device for indicating the speed should be installed on every car as the cost of one fine will purchase a reliable speedometer.

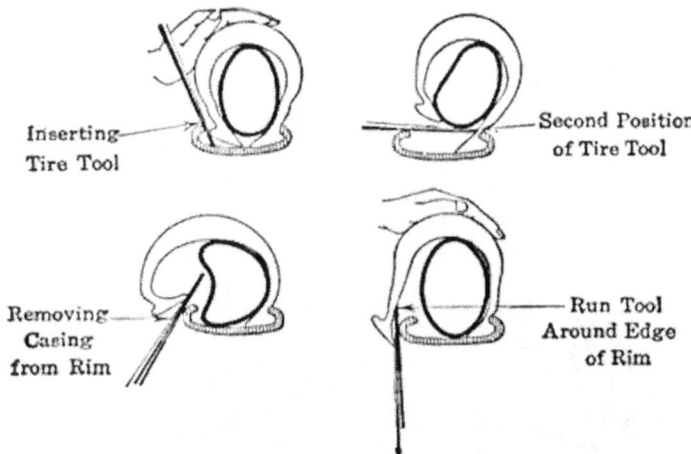

FIG. 29.—Method of removing clincher tires.

The drive may be taken from a gear attached to the transmission, as shown in Fig. 30, or from a similar attachment on one of the front wheels.

Figure 31 shows a speedometer drive installed in the spindle of the steering knuckle and driven from a plate under the hub cap. This eliminates the use of an exposed gear and requires no attention except proper lubrication. Care should be used to see that the drive plate is properly replaced if the hub cap is removed for any reason.

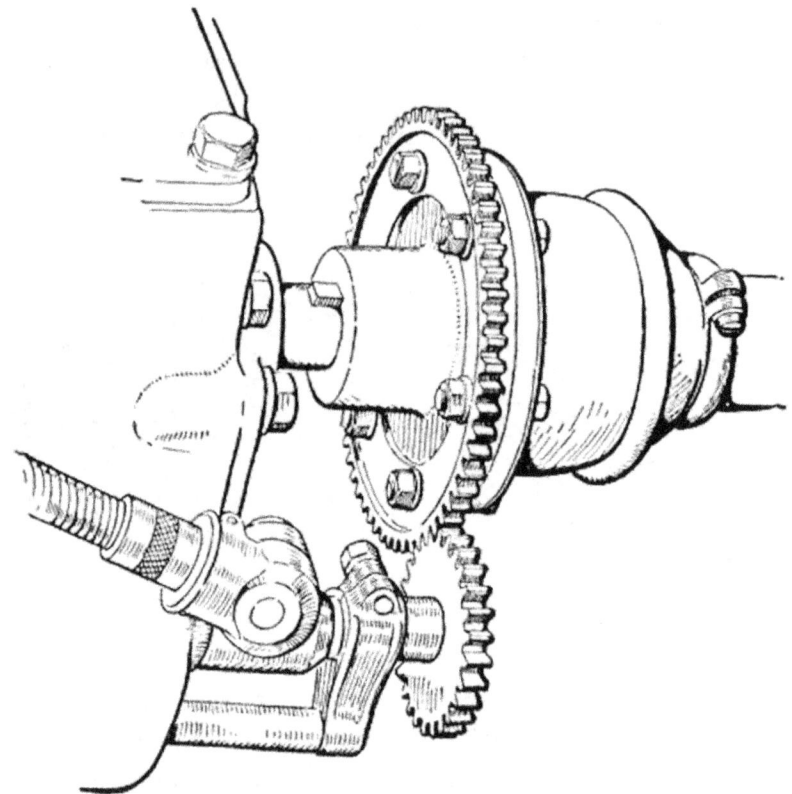

Fig. 30.—Speedometer drive from transmission.

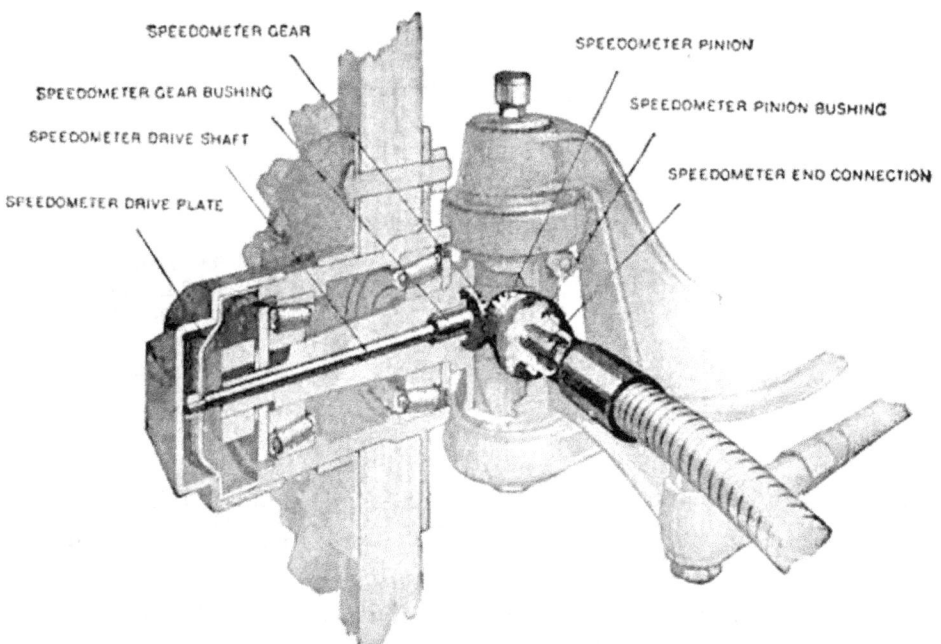

Fig. 31.—Speedometer drive through knuckle spindle.

20. Control Systems.—Figures 32 and 33 show the two prevailing control systems. Figure 32 shows the left-hand drive and center control system generally used on cars with sliding gear transmissions.

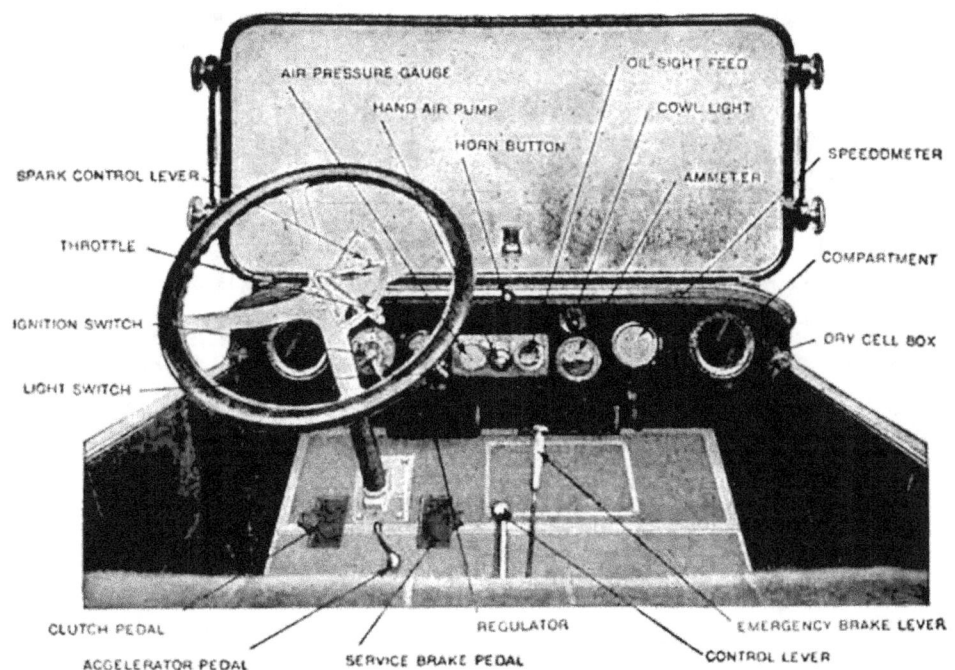

Fig. 32.—Left-hand drive, center control.

Fig. 33.—The Ford control.

The operation is as follows: The left-hand pedal operates the clutch and the other pedal the foot or service brake. The right-hand lever operates the emergency brake. The left-hand lever operates the change gears as follows: To the left and ahead for reverse, to the left and back for low speed ahead, to the right and ahead for second speed ahead, and to the right and back for third or high speed ahead. This order of events is not standard for all cars. Every car has its own system of shifting gears.

Figure 33 shows the Ford control system. This system consists of three foot pedals and one hand lever. The pedal on the left operates the clutch and controls the high and low speed. The hand lever also operates the clutch and when drawn all the way back sets the emergency brake. With the hand lever forward and left pedal up it is then in high gear. To get low speed ahead, the left pedal is pressed all the way forward; halfway in releases the clutch. The second or middle pedal marked "R" operates the reverse mechanism. To reverse the car the hand lever must be in a vertical position or the clutch pedal halfway in; then pressing on the reverse pedal drives the car backward. The right-hand pedal operates the foot or service brake, which is on the transmission.

The chapters to follow will treat in detail of the various parts of the car, their construction, methods of operation, and maintenance.

CHAPTER II

ENGINES

21. What is an Explosion?—Practically all gasoline engines are driven by explosions which take place within the cylinder of the engine and drive the piston, thus causing rotation of the revolving parts of the engine. These explosions are in a way very similar to the explosions of gunpowder or dynamite. When a charge of gunpowder is fired in a cannon or gun, the gunpowder burns and produces gases which exert a tremendous pressure on the shell and force it from the gun.

Practically any substance that will burn can be exploded if under the proper conditions. An explosion is merely a burning of some material taking place almost instantaneously, so that a great amount of heat is generated all at once. When any substance burns, it unites rapidly with oxygen from the air. If we want to get an explosion, it is necessary to have the fuel very finely divided and carefully mixed with air, so that the burning can be very rapid. Then, if we start the fuel burning, by an electric spark or any other means, the flame instantly spreads throughout the mixture and an explosion occurs. In a gasoline engine we take in gasoline vapor mixed carefully with air. This mixture is then exploded inside the cylinder of the engine. The force of this explosion drives the piston and the motion is transmitted through the connecting rod to the crank. To make the process continuous and keep the engine going, it is necessary to get rid automatically of the gases from the previous explosion and to get a fresh charge into the cylinder ready for the next explosion. This process must be carried out regularly by the engine, in order to keep it running.

22. Cycles.—As we have just seen, an engine must supply itself with an explosive mixture so that the force of the explosion will cause the engine to move, and it must get rid of these dead gases and get in a fresh charge of gas and air and explode this so as to keep up the motion. There are in use at the present time two principal systems of performing this series of operations. These systems, or rather the series of operations, are called *cycles*, and the engines are named according to the number of strokes it takes to complete a cycle. These two cycles, or systems of engines, are the *four-stroke cycle* and the *two-stroke cycle*.

Remember that a *cycle* refers to the series of operations the engine goes through. In the four-stroke cycle there are four strokes or two revolutions. In the two-stroke cycle there are two strokes or one revolution. Many people leave out the word *stroke* and talk of "four-cycle engines" and "two-cycle engines." This causes the misunderstanding that many people have as to just what a cycle really is. A better way is to call them "four-stroke engines" and "two-stroke engines."

23. The Four-stroke Cycle.—Figures 34, 35, 36 and 37 show an engine which operates according to the four-stroke cycle. The engine shown here is a vertical engine, that is, the cylinder is placed above the crank shaft (instead of being at one side) and the piston moves up and down in the cylinder. This is the prevailing form for automobile engines.

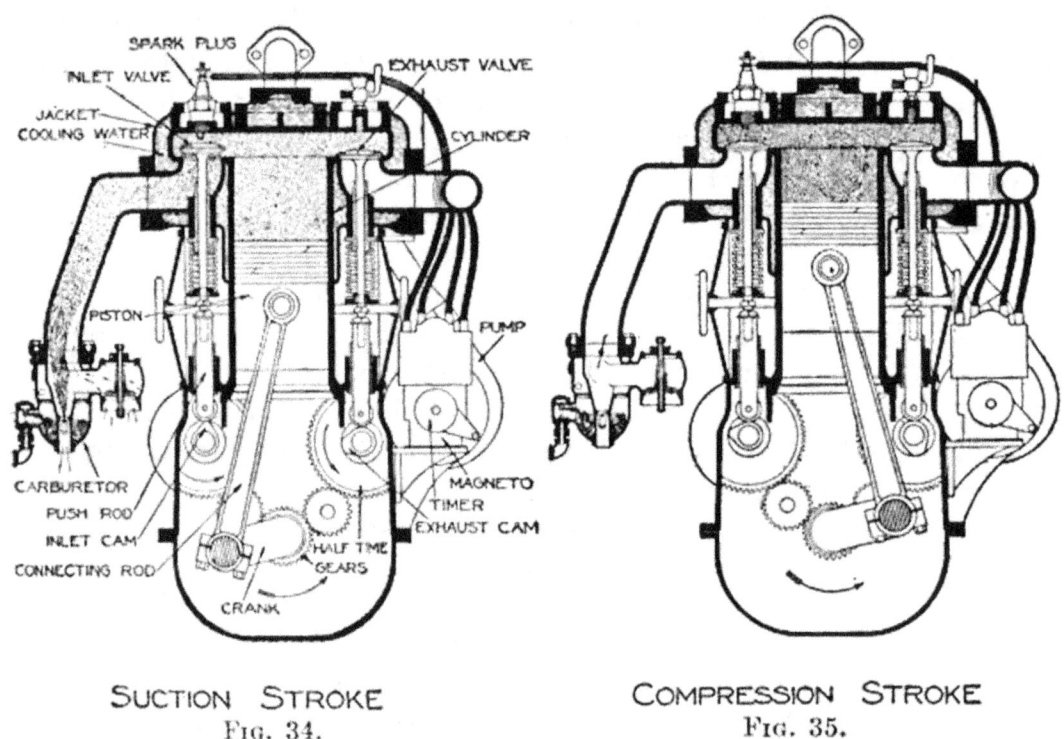

SUCTION STROKE
FIG. 34.

COMPRESSION STROKE
FIG. 35.

Any engine consists of four principal parts: the *cylinder*, which is stationary and in which the explosion occurs; the *piston*, which slides within the cylinder and receives the force of the explosion; the *connecting rod*, which takes the force from the piston and transmits it to the crank; and lastly the *crank*, which revolves and receives the force of the explosion as the piston goes in one direction, and which then shoves the piston back to its starting point. A four-stroke engine has a number of other minor parts, whose uses will be brought out presently. This engine uses four strokes of the piston to complete the series of operations from one explosion to the next, and is therefore said to operate on the four-stroke cycle, or it is said to be a "four-stroke" engine. The first illustration, Fig. 34, shows the engine just drawing in a mixture of gas and air. This is continued until the piston gets clear down to the bottom of the stroke, and the cylinder is full of this explosive mixture. This operation is called the *suction stroke*. Then the valves are shut, as in Fig. 35, and the piston is forced back to its top position. This squeezes or compresses the gas into a space left in the top of the cylinder, and this process of compressing it is called the *compression stroke*. After the piston gets to the top, the gases are ignited or set fire to and burn so quickly that an explosion results and the piston is driven down again, as in Fig. 36. This is called the *expansion or working stroke*. When it reaches the bottom of the stroke, another valve is opened, and while the piston is returning to the

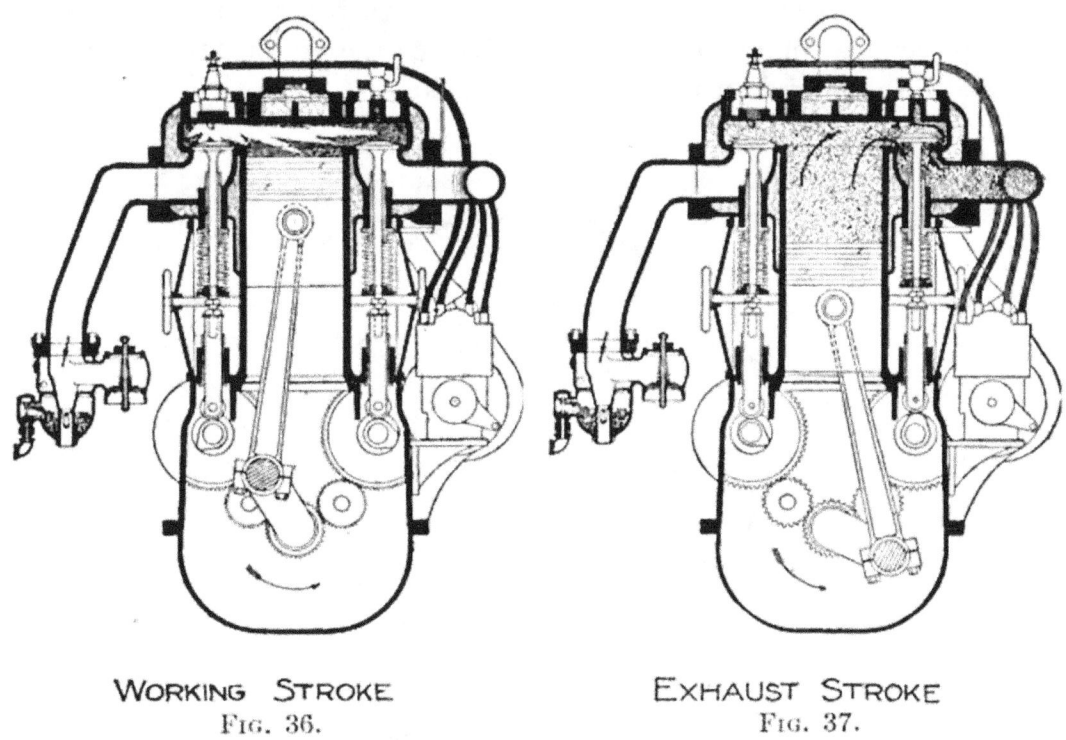

WORKING STROKE   EXHAUST STROKE
Fig. 36.   Fig. 37.

top position it forces out through this valve the burned gases which occupy the cylinder space. This is the *exhaust stroke*. The engine is now ready to repeat this series of operations. These operations have taken two revolutions or four strokes. A stroke means a motion of the piston from either end of the cylinder to the other end. Consequently, there are four strokes in the cycle of operations of this engine, and we therefore call it a four-stroke engine.

24. The Order of Events in Four-stroke Engines.—The various parts or events in the four-stroke cycle are shown on the diagram of Fig. 38. This shows the two revolutions of the four-stroke cycle divided up so as to show the crank positions when the different events occur. The diagram is drawn for a vertical engine with the crank revolving to the left, as shown on the engine of Figs. 34 to 37. This is the direction of rotation of an automobile engine to a person in the car looking forward toward the engine.

Starting at the top of the diagram, we have just exploded the charge and as the crank swings over to the left the gases are expanded. Before the crank reaches the bottom, the exhaust valve is opened. This is kept open while the piston is returned to the top. The inlet valve is then opened and the suction stroke occurs as the crank and piston again descend. Just after the crank passes the bottom, the inlet valve closes. Both valves being now closed, the charge is compressed as the crank and piston rise again to the top. A short time before reaching the top, ignition occurs. This should be just far enough before the top so that the explosion or combustion is taking place as the crank passes the top and starts to descend on the expansion stroke.

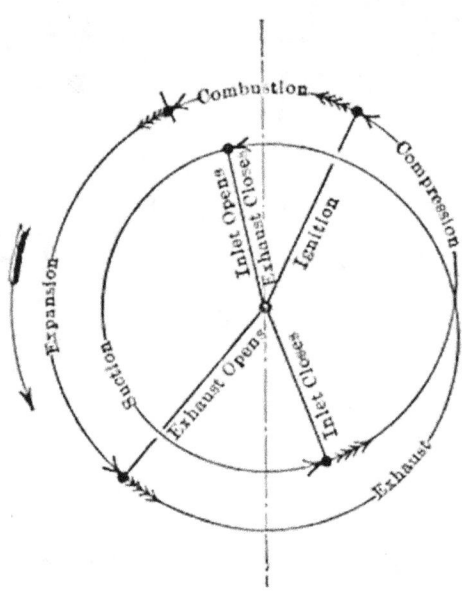

Fig. 38.—Order of events in the four-stroke cycle.

25. The Mechanism of Four-stroke Engines.—In addition to the four principal parts previously mentioned, there are a number of other small parts which we will now discuss. First, we must have two valves located in the upper end of the cylinder, one for the purpose of letting in the fresh mixture of gas and air, and the other for the purpose of letting out the burned gases. Each of these valves opens once in a cycle, that is, once in two revolutions. In this engine (Figs. 34 to 37) the valves are shown in the T-head arrangement, the inlet valve being on the left and the exhaust valve on the right. These valves are of a form called *poppet* valves. They are mushroom shaped, with beveled edges which fit into a beveled seat. The valves are held shut by springs on the outside, which pull on the valve stems and hold them tightly against the seat, so that gases can not leak in or out, except when one of the valves is opened. To operate the valves, there are two push rods, one for each valve. These push rods receive their motion from the cams. On the lower ends of these rods are rollers, and these roll on cams on the cam shaft inside of the crank case. These cams have each a hump or projection on about one-fourth of their circumference. When one of these strikes the roller it raises it up, and this motion is transmitted through the push rod to the valve. After the projection of the cam has passed under the roller, the valve spring will close the valve and force the push rod back to the original position.

Since the valves on an engine each work but once in two revolutions, the engine must be arranged so that the cams come around only once in two revolutions. To do this, the general arrangement is to put a small gear on the crank shaft and have this drive another gear, twice as large, on the cam shaft. In this way the cam shaft will run at just half the speed of the crank shaft. These gears are called *half-time* gears.

26. Valve Timing and Setting.—The exhaust valve of an engine opens on an average of about 45° before the end of the stroke, in order that the pressure may be reduced to atmospheric by the end of the stroke so there will be no back pressure during the exhaust stroke. At the end of the exhaust stroke, the exhaust valve should remain open while the crank is passing the center so that any pressure remaining in the cylinder may have time to be reduced to atmospheric.

The inlet valve very seldom opens before the exhaust closes. Most manufacturers do not open the inlet until the exhaust closes, for fear of back-firing, although there is little danger of this except with slow-burning mixtures. The inlet valve opens, on an average, 10° late (after center). At the end of the suction stroke there is still a slight vacuum in the cylinder and the inlet is kept open for a few degrees past center to allow this to fill up and get the greatest possible quantity of gas into the cylinder. On an average, the inlet valve closes about 35° late, depending on the piston speed of the engine.

In studying the valve setting of an engine, the first step, of course, is to observe the timing of the engine as it stands. To do this we must turn the engine by hand. By inserting a thin sheet of tissue paper between a valve stem and its push rod, we can tell when the valve opens and closes by noticing when the paper is gripped in opening the valve and when it is released in closing. The corresponding crank positions should be noted. We can then see whether it is possible to do anything to improve the valve setting. Valve cams are made for a certain valve setting and will give a certain angle of opening. This may become altered in several ways. Any excessive lost motion in the valve motion will result in a valve's opening too late and closing too early. Wear on the cam will have the same effect. If a cam shaft has been removed and replaced, the timing gears may be put together wrong. This would advance or retard the whole series of events and can readily be found out when the timing is observed.

The clearance or lost motion in the valve mechanism between the cam and the valve stem should be about $\frac{1}{64}$ in. or less. In order to keep the valves quiet on their engines, some makers use a clearance of the thickness of ordinary writing paper, or about $\frac{5}{1000}$ in. If the clearance or lost motion is too great, it will cause the valve to open late and close early, and will also cause the cam to strike the roller a hard blow with the middle of its face, instead of catching it gradually at the beginning of the incline. It will also reduce the valve opening and possibly choke the engine.

In a four-stroke engine the cam shaft revolves once for each two revolutions of the crank shaft. Consequently, a valve opening of 180° will be represented by but 90° on the cam, and, for any given crank angle through which the valve is to be open, the corresponding cam angle will be but one-half the given crank angle. If an exhaust valve is to open 45° before the beginning of the exhaust stroke and close 10° after the end of the stroke, the total crank angle will be

$$180° + 45° + 10° = 235°$$

The corresponding cam angle $= \frac{235°}{2} = 117\frac{1}{2}°$. By "cam angle" we mean the angle on the cam, from the point where it starts to open the valve to the point where the valve is seated again. An inlet valve that is to open 10° late and close 30° late, would have a total crank angle of

$$180° - 10° + 30° = 200°$$

The corresponding cam angle $= \frac{200°}{2} = 100°$.

27. Valves.—The prevailing type of valve is what is called the *poppet* or *mushroom* type—poppet, from its operation, and mushroom, from its shape. The exhaust valve must be opened by a cam because it must be opened against a pressure of 40 to 60 lb. in the cylinder and held open while gases are forced out through it. The inlet valve may be opened by a cam or we may use a light spring and depend on the suction to open it. The suction type is, of course, cheaper to build, but it reduces the capacity of the engine so that for the same power there is no saving. Consequently we find automatic inlets as a rule only on the small farm engines that are built to sell at a low price. To open an automatic valve, there must be a difference in pressure on the two sides of the valve equal to the tension of the valve spring. This tension may be reduced or increased by the weight of the valve, if vertical, and opening respectively downward or upward. For high-speed engines an automatic valve is particularly unsuited, since a heavy spring must be used to insure quick closing at high speed.

Poppet valves usually have 45° beveled seats as shown in Fig. 39, though occasionally flat valves are seen which rest on flat seats. The valves must be large enough to let the gases in and out of the cylinders freely. If they are too small they will cut down the power of the engine by not permitting it to get a full charge. The valves usually measure from one-third to one-half of the cylinder diameter. Valve diameters are usually measured by the opening in the valve seat (see dimension marked d in Fig. 39). The diameters of the inlet and exhaust pipes should at least equal this valve diameter and should be larger if possible.

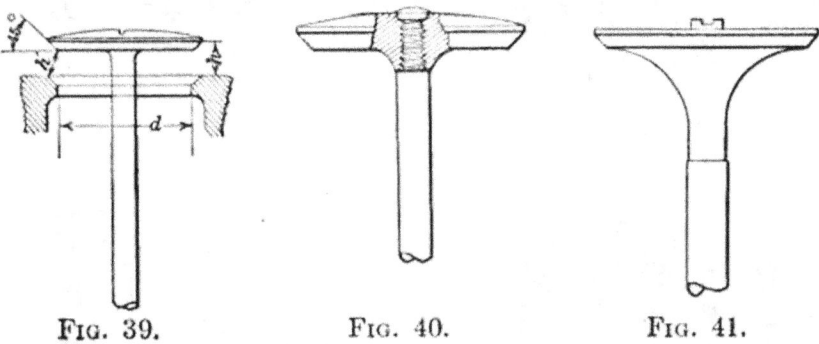

Fig. 39.   Fig. 40.   Fig. 41.

The valve lift should, when possible, be sufficient to give the gases as large a passage between the valve and seat as they have through the opening d, Fig. 39. For a flat valve seat this would require a lift of one-fourth of the valve diameter. With a beveled seat, the gases pass through an opening in the shape of a conical ring having a width of passage equal to h', Fig. 39. To have the necessary passage area, the lift h of the valve should be about three-tenths of the diameter. In most stationary engines this lift can be given the valve, but in high-speed engines it would be too noisy. This lift would then cause pounding and wear on the cams; it would require very stiff springs to make the valves follow the cams in closing and would be very hard on the valve seats and stems. For automobile engines the valves are made as large as possible and the lift is limited to from $5/16$ to $1/2$ in.

The best materials for valve heads are cast-iron, nickel-steel, and tungsten-steel. Cast-iron is very cheap, easily worked, and stands corrosion well. It is weak, however, and therefore requires a heavier weight than other materials and this is especially objectionable for high-speed engines. The nickel-steel is strong, non-corrosive, and has a very low coefficient of heat expansion. Hence it does not warp so readily as other metals It is rather expensive and when used is generally electrically welded to a carbon-steel valve stem. The tungsten-steel is very hard and will stand high temperatures without pitting. Cast-iron valve heads can be screwed on a steel stem as in Fig. 40, the stem being riveted to prevent loosening. Figure 41 shows a common European form

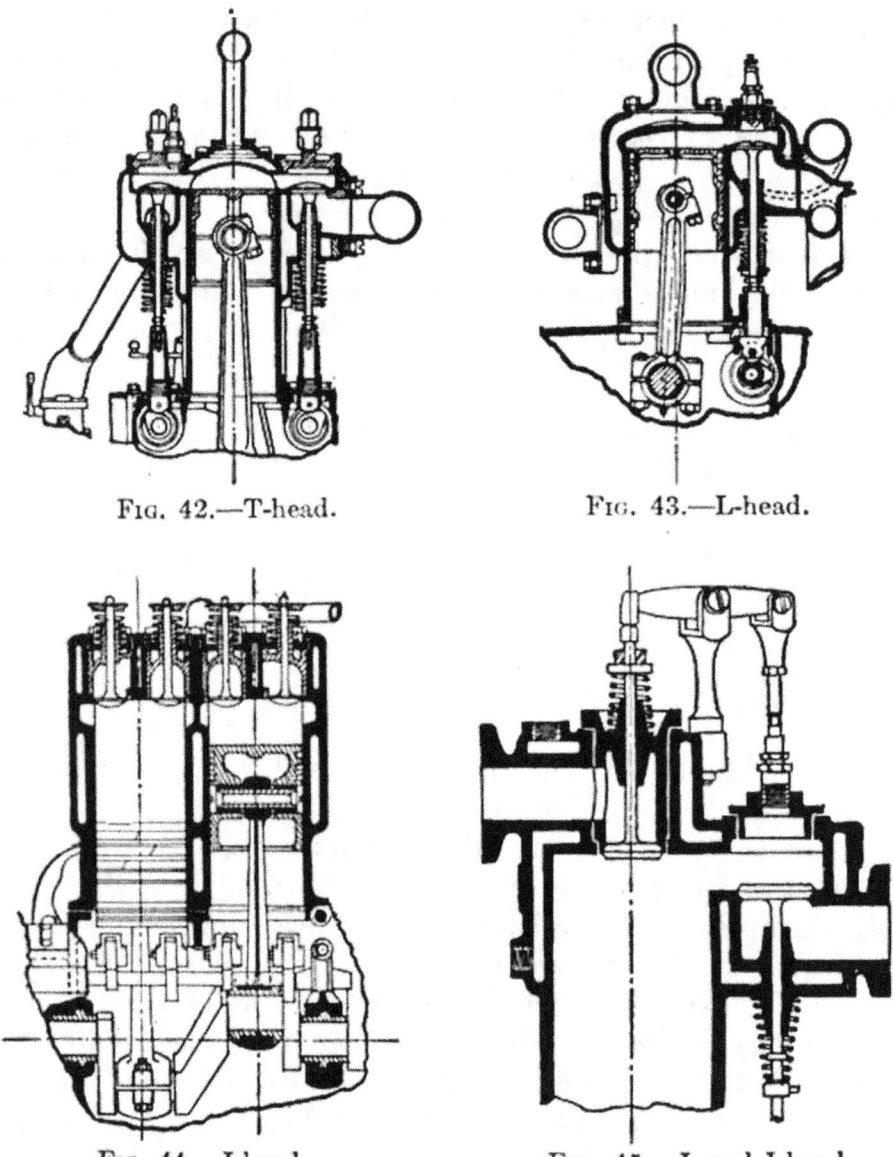

FIG. 42.—T-head.   FIG. 43.—L-head.

FIG. 44.—I-head.   FIG. 45.—L-and-I head.

for valves which is being rapidly adopted here. The curvature underneath gives the gases a smooth passage without any of the whirling eddies that occur under the ordinary flat valve.

Any valve needs regrinding into its seat occasionally with oil and emery or ground glass. Exhaust valves require this more often than inlet valves, as they become warped and pitted by the hot gases. After a valve is ground in, the push rods should be readjusted, as the grinding will lower the valve and reduce the clearance in the valve motion.

28. Valve Arrangements.—The possible arrangements of the valves in the cylinder are numerous. Figure 42 shows the T-head arrangement used in many of the large automobiles. This arrangement permits of a large valve and a low lift, and therefore makes a very quiet engine. Figure 43 shows the L-type with both valves on one side. This is the most common type. It requires only one cam shaft and has a very simple,

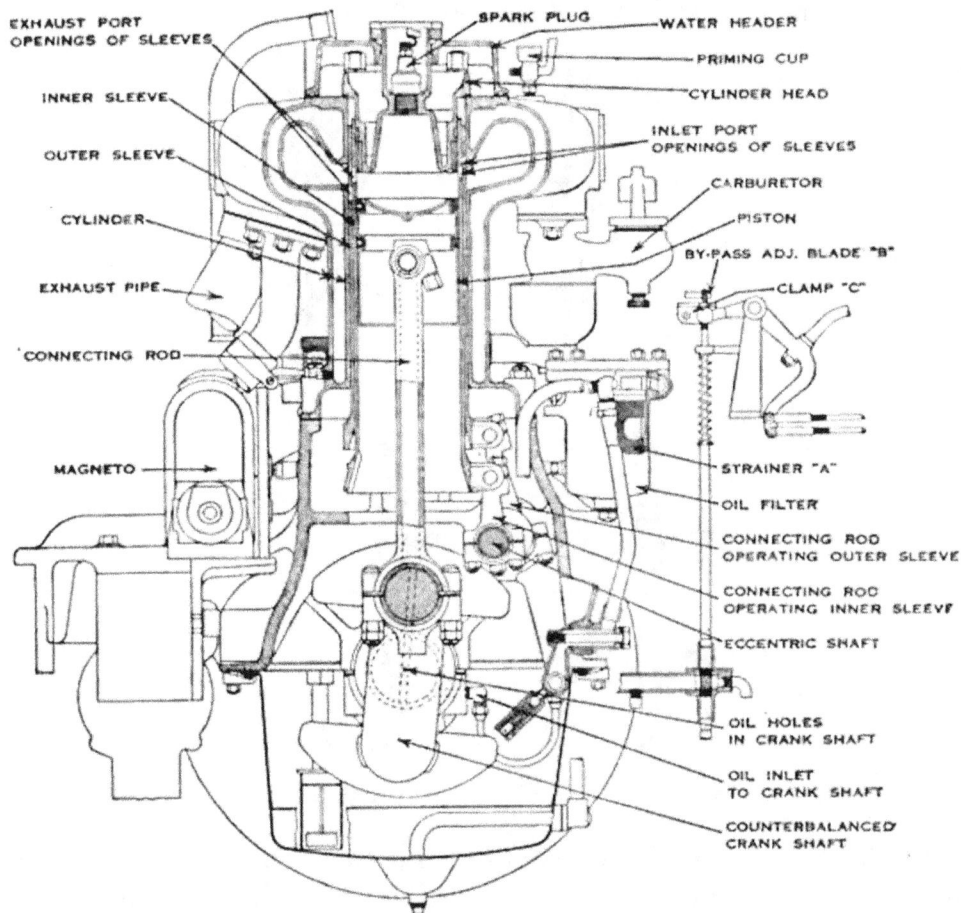

FIG. 46.—Section of Silent Knight engine.

direct-acting valve mechanism. It does not have as much cooling surface to the combustion chamber and is, therefore, more economical in the use of fuel than the T-head. Figure 44 shows the valve-in-the-head arrangement. This is sometimes called the I-head arrangement. It is especially popular for racing cars because it gives a short, quick passage into the combustion chamber and gives a simple, compact combustion chamber with a minimum loss of heat to the cooling water. Figure 45 shows an arrangement used on the Reo car that is a combination of the L-type and the valve-in-the-head type, the intake valve being in the top and operated by a rocker arm while the exhaust is on the side and is operated by a direct push rod. Both valves are operated from one cam shaft.

29. The Knight Engine.—The Knight engine is built on the principle of the four-stroke cycle, but the usual poppet valves have been replaced by two concentric sleeves sliding up and down between the piston and cylinder walls. Certain slots in these sleeves register with one another at proper intervals, producing direct openings into the combustion chamber from the exhaust and inlet ports. The construction of the Stearns-Knight motor is illustrated in Fig. 46 which shows the general arrangement of the parts and their nomenclature.

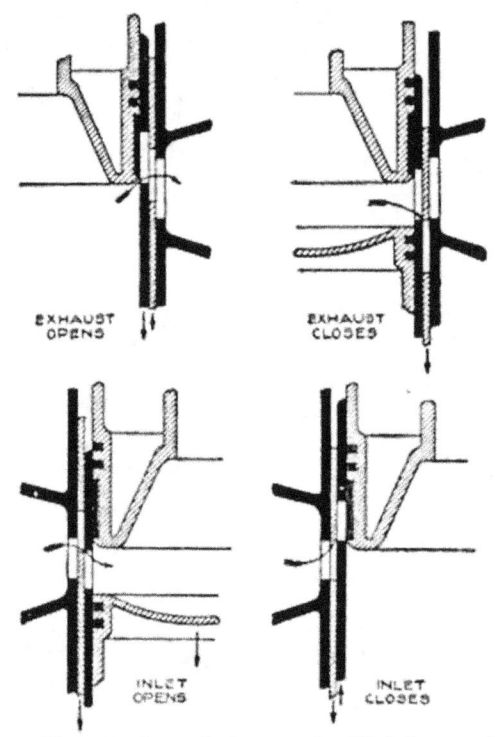

Fig. 47.—Action of sleeves in Knight engine.

It will be noted that two sleeves are independently operated by small connecting rods working from an eccentric or small crank shaft running lengthwise of the motor. This eccentric shaft is positively driven by a silent chain at one-half the speed of the crank shaft. The eccentric pin operating the inner sleeve is given a certain lead or advance over that operating the outer sleeve. This lead, together with the rotation of the eccentric shaft at half the crank-shaft speed, produces the valve action illustrated in Fig. 47, which shows the relative positions of the piston, sleeves, and cylinder ports at various points in the rotation of the crank shaft.

30. The Rotary Valve.—The rotary valve as used in the Speedwell car consists of two cylindrical shafts in the head of the motor, one for exhaust and one for the inlet. These shafts are slotted and when rotating register with ports in the cylinder walls, thus opening passageways for intake and exhaust gases. The rotary movement of the valves is continuous in one direction, the valves being driven by a silent chain from the crank shaft. Figure 48 illustrates the different positions of the rotary valves at the beginning of each of the four strokes. The arrows inside show the direction of rotation of the valves and the arrows outside indicate the direction of the fresh gas going in and the exhaust gas passing out of the cylinder.

31. Two-stroke Engines.—Two-stroke engines as a class are not so flexible as the four-stroke engines under the varying speeds and loads encountered in automobile service. Consequently they have not been used to any great extent in motor cars, although a few satisfactory cars have been built with them.

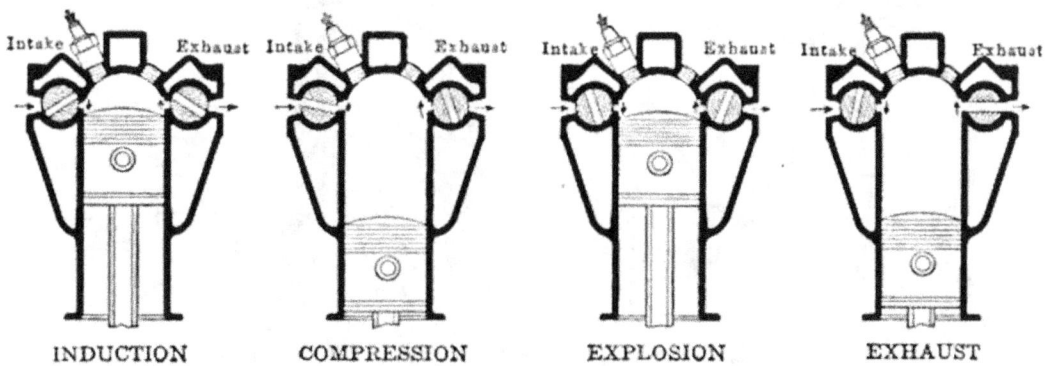

FIG. 48.—Speedwell rotary valve engine.

Since the piston of a four-stroke engine receives an impulse or explosion only once in two revolutions, considerable effort has been expended in trying to develop an automobile engine that would give an explosion in each cylinder every revolution and yet would operate as satisfactorily and economically as the four-stroke engine. An impulse every revolution would make a more powerful engine than one of the same size which received an impulse only once in two revolutions and it would also make the flow of power more continuous for the same number of cylinders.

The Two-port Engine.—Most of the two-stroke engines in use are very much like those shown in Figs. 49 to 52. In appearance, these engines are much simpler than the four-stroke engine, but are not necessarily any simpler in operation. They do not have any valves opening into the combustion chamber, such as are found in the four-stroke engine. The exhaust gases leave the cylinder through a port in the cylinder wall, which is uncovered by the piston at the end of the expansion stroke, as shown in Fig. 50. At the same time, a fresh charge is blown into the cylinder through a similar port on the other side. The top of the piston has a deflector which turns the incoming charge up into the clearance space. The charge then strikes the cylinder head, which turns it down on the other side toward the exhaust port, thus driving the dead gases out ahead of it. The piston then comes back, shuts off both these openings and compresses the fresh charge into the clearance space as shown in Fig. 49. It is then ignited in the usual manner by a spark plug screwed into the cylinder head. This gives the piston an impulse every revolution.

The engines of Figs. 49 to 52 have each crank enclosed all around and they use this case or chamber as a sort of a pump to supply fresh gas to the cylinder. When the piston goes up, the space inside the crank case is increased, and when it comes down the space is reduced, thus maintaining a breathing action inside the crank case. In Fig. 49 the piston

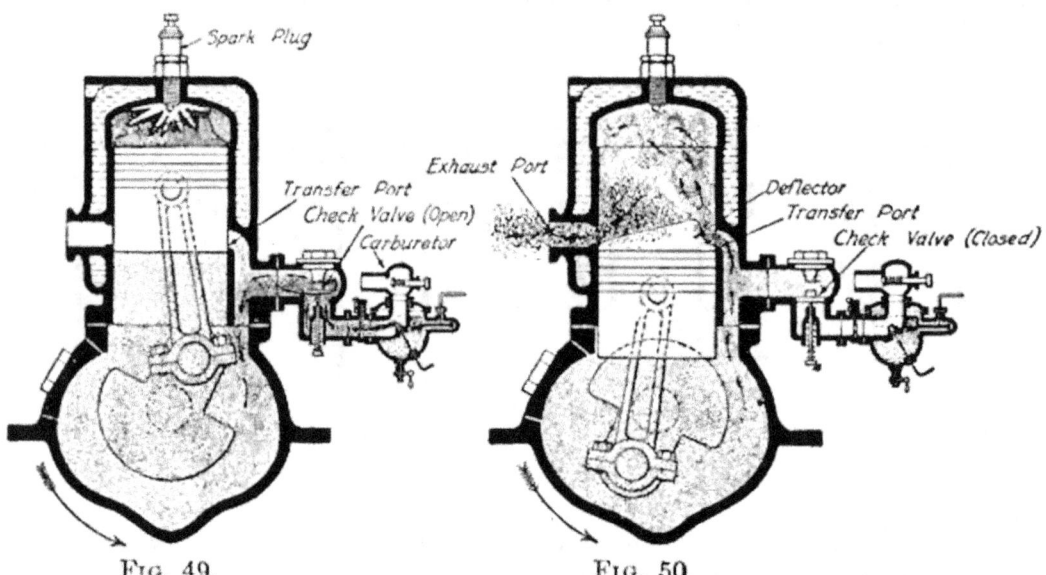

Fig. 49. Fig. 50

Figs. 49 and 50.—Two-port, two-stroke engine.

is shown traveling toward the top. This motion causes a suction in the crank case and causes air to enter through the carburetor. As the air passes through the carburetor it becomes saturated with gasoline and then passes through the check valve into the crank case. When the piston gets to the top, the suction ceases and the check valve is closed by its spring. Meanwhile, an explosive mixture has been compressed above the piston and at the top of the stroke is ignited by a spark. This produces an explosion or rise in pressure above the piston, just as in the four-stroke cycle and this drives the piston down on its working stroke.

As the piston comes down, it compresses the fresh gases in the crank case into a smaller volume and thus raises their pressure. Meanwhile, as the piston nears the bottom of its stroke, it uncovers the exhaust port and the pressure in the cylinder causes a large part of the burned gases to shoot out through this port. An instant later the piston uncovers a transfer port on the other side and is now in the position shown in Fig. 50. This transfer port is connected into the crank case and therefore allows the gases from the crank case to blow over into the cylinder as shown in Fig. 50.

The piston head is so shaped as to form a deflector, which turns the fresh charge toward the cylinder head so that it can not blow out the exhaust port. The piston then returns, cuts off these ports, and compresses this charge, meanwhile drawing another charge into the crank case. This engine is called a *two-port* type, because there are only two ports in the cylinder walls to be operated by the piston.

The Three-port Engine.—The only difference between this type and the preceding one is in the method of admitting the gases into the crank

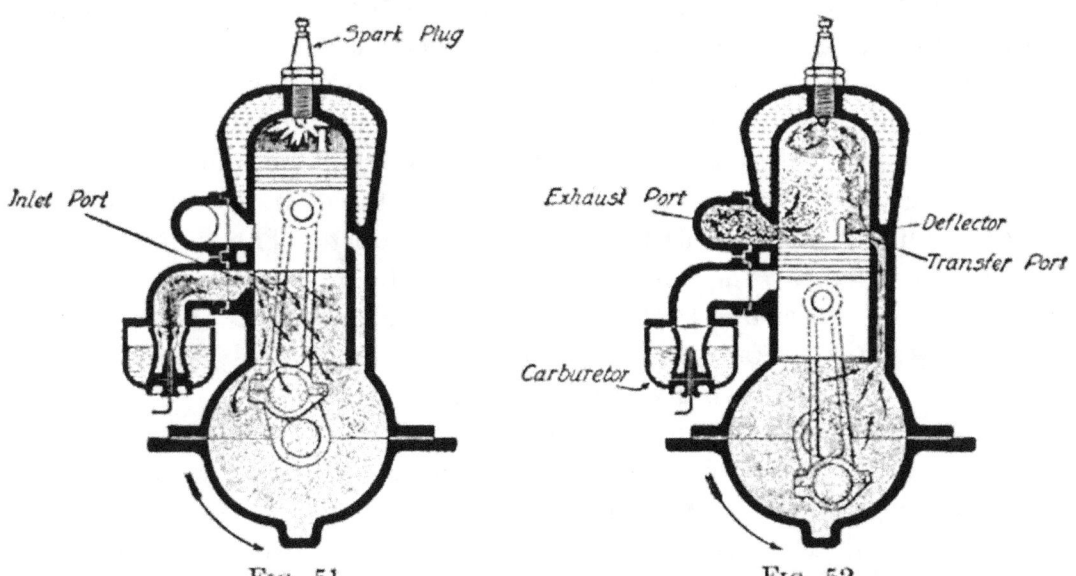

FIG. 51.   FIG. 52.
FIGS. 51 AND 52.—Three-port, two-stroke engine.

case. Instead of using a check-valve, the admission of the gases to the crank case is controlled by the piston, which uncovers a *third* port in the cylinder walls as it nears the top of the compression stroke. As will be seen in Fig. 51, the carburetor is on the other side of the engine, placed just below the exhaust pipe. As the piston rises, it creates a suction in the crank case, but there is no way for any gas to get in until the piston reaches the top of its stroke. As the piston uncovers this third port, the air enters with a rush through the carburetor, picks up the gasoline on its way through, and enters the crank case. The piston then descends, cuts off the third port, compresses the gases in the crank case, as in Fig. 52, and then blows them over into the cylinder as before.

Against the two-stroke engine we have the facts found from experience that they are not as economical in the use of fuel and are more uncertain in their action than the four-stroke engine. Since the fresh charge is depended on to blow out the exhaust gases, it is evident that some of the incoming charge is liable to pass out through the exhaust port. Gases mix very quickly and it is not possible to keep the dead and fresh gases separate, and yet drive the dead gases out and fill the cylinder completely with fresh gases. If a full charge enters through the transfer port, some of it will be lost through the exhaust port without its being utilized. By skillfully proportioning the two ports and the shape of the deflector to the size and speed of the engine, it is possible to largely prevent the waste of fuel through the exhaust port.

A two-stroke engine does not get as full a charge of gas as does a four-stroke engine and, consequently, will not be twice as powerful. The horse power of a two-stroke engine is usually about $1\frac{1}{3}$ to $1\frac{1}{2}$ times that of a four-stroke engine of the same size and speed.

The small two-stroke engines shown in Figs. 49 to 52 sometimes cause trouble from back-firing or exploding in the crank case. This is caused by the mixture in the crank case becoming ignited and exploding before it goes over into the cylinder. This wastes the energy of the gas and fills the crank case with dead gases, so that the engine will frequently come to a stop. Back-firing is caused by the mixture in the cylinder being still in flames when the piston uncovers the transfer port. The flame shoots through this port into the crank case and fires the mixture there. It has been found by experience that mixtures weak in gas are the ones which burn slowly and therefore cause back-firing. Consequently, the cure for crank-case explosions is to give the engine more fuel.

Any leaks into the crank case are very serious in either of these types. With the slow speed used in starting an engine by hand, a very small leak may admit air enough to satisfy the suction in the crank case and thus prevent any gas from being drawn in or, at any rate, it may so weaken the mixture as to make it non-explosive.

This brief statement of some of the difficulties of the two-stroke engine will show some of the things that must be overcome in order to make this type of motor generally applicable to automobile service.

32. The Flywheel.—The purpose of the flywheel is to keep the engine running from one explosion to the next, and to make the engine run smoothly. If an engine did not have a flywheel, it would run in a very jerky manner, if it ran at all, and it is more probable that the explosion would simply drive the piston to the other end of the stroke and that it would stop there. Any one knows that the heavier a moving object is and the faster it is going, the harder it is to stop it. The flywheel on an engine is quite heavy and the result is that, once started, it will keep the engine going for some time. A gas-engine flywheel must not only be heavy enough to keep it going from one explosion to the next, but must keep it going without allowing the speed of the engine to drop down too much between explosions.

33. Ignition.—In order to cause the explosions within the cylinder, some means must be provided for lighting the charge of gas. This is usually done by causing an electric spark to pass between two points within the cylinder. The spark sets fire to the mixture and the explosion follows.

There are two general methods of electric ignition. One of these is called the *make-and-break system* because it requires some moving parts inside the cylinder to *make* an electric circuit, and then *break* it quickly so that a spark will occur inside the cylinder. The other system is called the *jump-spark system*. This is the system used in automobiles. There are no moving parts which have to pass through the cylinder wall in this system. The spark coil or magneto makes a current powerful enough to jump between two fixed points inside the cylinder. The complete details of these systems of ignition will be taken up in a later chapter.

34. Clearance and Compression.—It was discovered by some of the early inventors of gas engines that compressing a gaseous mixture causes it to give a much more powerful explosion. Consequently, all gas engines draw in a full cylinder charge of gas and air, and then compress this back into a space left at the upper or rear end of the cylinder. This space, which is left for the gas to occupy when the piston is at the top end of its stroke, is called the *clearance space* or *combustion chamber*. The amount of this clearance space in relation to the whole cylinder volume determines just how much the gas is compressed. It has been found from experience that different kinds of gases require different amounts of compression and, therefore, the clearance space is made different for different fuels. The clearance is generally spoken of as being a certain *per cent.* of the piston displacement, varying from 24 to 30 per cent. for automobile engines.

35. Piston Displacement.—This refers to the space swept through by the piston in going from one end of the stroke to the other. It is given this name because, as the piston moves through its stroke, it will either draw in or force out that volume of air or gas. The piston displacement is calculated by multiplying the length of stroke by the area of a circle whose diameter is the inside diameter of the cylinder. For example, a $3\frac{1}{2}$-in by 5-in. engine (this means $3\frac{1}{2}$ in. inside cylinder diameter and 5 in. stroke) would have a piston displacement as follows:

The area of a $3\frac{1}{2}$-in. circle is $0.7854 \times 3\frac{1}{2} \times 3\frac{1}{2} = 9.621$ sq. in.

The piston displacement is 5 times this, or 48.105 cu. in.

The clearance of such an engine would be from 24 to 30 per cent. of this. If we suppose that it is 25 per cent., then the actual space which must be left for the clearance will be $48.105 \times 0.25 = 12.026$ cu. in.

36. Cylinder Cooling.—When an explosion occurs inside the cylinder of an engine, the gases on the inside reach a temperature somewhere around 3000°. The walls of the cylinder are, of course, exposed to this high heat and would very quickly get red hot if we did not have some way of keeping them cool. The polished surface upon which the piston slides would be very quickly spoiled. The most common way of keeping the cylinder cool is by the use of water, and the arrangement for this is shown in the engines illustrated in this chapter. Surrounding the cylinder is a jacket with a space between for the cooling water. By keeping a supply of water passing through this space, the cylinder can be kept cool enough for the operation of the engine. The cylinder head is also cast with a double wall, especially around the valves, so that these parts will also be kept cool. The cooling fluid used is generally water, although sometimes special anti-freezing solutions are used where there is danger of the engine freezing. Water should not be allowed to remain in the jacket of an engine over night if there is danger of a frost, as the freezing of the water will crack the cylinder. When the supply of water is limited, as in an automobile, the water is cooled in a radiator or system of pipes, and used over again. The water is kept in circulation by a pump or by the thermo-syphon system and the hot water is cooled by the air passing over the radiator.

37. The Muffler.—When the exhaust valve of an engine opens at the end of the expansion stroke the pressure of the gas inside the cylinder is

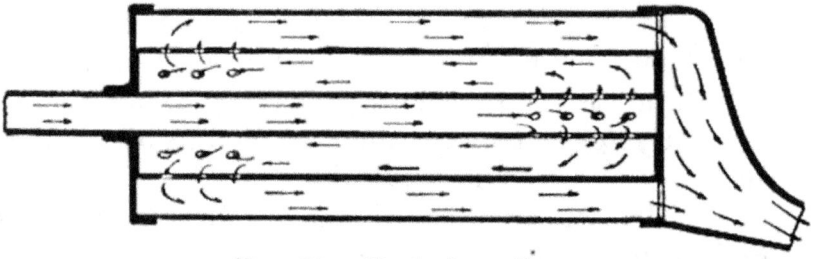

Fig. 53.—Typical muffler.

still about 50 or 60 lb. per square inch. The valve must open and let this pressure out before the piston starts back, or else the back pressure will tend to stop the engine. The valve is opened quickly, and the high pressure, being suddenly released into the exhaust pipe, causes the sharp sound which we hear when an engine exhausts. This sound is not the sound of the explosion, as is commonly supposed. The real explosion takes place a little before this sound and can be heard only as a dull thump inside the cylinder. The explosion occurs at the *beginning* of the working stroke, while the sound that we hear in the exhaust comes at the *end* of the stroke.

In order to prevent this sudden exhaust from causing too great a noise it is customary to have a muffler. A muffler is generally a chamber in the exhaust pipe which receives the exhaust gases from the engine and expands them gradually into the outside air, thus preventing a loud noise. A common arrangement of an automobile muffler is shown in Fig. 53.

38. Horse Power of Engines.—The horse power of an engine is the measure of the rate at which it can do work. One horse power is a rate of 33,000 ft.-lb. a minute. There are two ways of measuring engine power. We can determine the power developed by the explosions in the cylinder, in which case we have what is called the *indicated horse power* (*i.hp.*); or we can attach a brake to the flywheel and measure the power which the engine actually delivers. This is called the *brake horse power* (*b.hp.*). Engines are usually rated by their brake horse power because that is what they are actually capable of delivering. The brake horse power of an automobile engine will usually be from 70 to 85 per cent. of its indicated horse power, the loss being that consumed in the engine mechanism.

There are a number of quick rules for estimating the power of engines according to their cylinder dimensions and the speed. Those most used for four-stroke engines are given below. The simplest of these and the one most used is known as the S. A. E. formula or Society of Automobile Engineers formula.

Authority	Formula
S. A. E. } Royal Auto Club	$\dfrac{D^2 N}{2.5}$ = hp.
Brit. Inst. of Auto Engrs.	$0.45 (D + L)(D - 1.18)$ = hp.
E. W. Roberts	$\dfrac{D^2 L R N}{18,000}$ = hp.

D = diameter of cylinder in inches.
L = length of stroke in inches.
R = revolutions per minute of crank shaft.
N = number of cylinders.

Derivation of the S. A. E. Horse Power Formula.—The indicated horse power of a single-cylinder, four-stroke engine is equal to the mean effective pressure, P, acting throughout the working stroke, times the area of the piston, A, in square inches, times one-quarter times the piston speed, S, divided by 33,000, thus:

$$\text{i.hp.} = \frac{PAS}{33,000 \times 4}$$

Multiplying this by the number of cylinders, N, gives the indicated horse power for an engine of the given number of cylinders, and further multiplying by the mechanical efficiency of the engine, E, gives the brake horse power.

Therefore, the complete equation for brake horse power reads:

$$\text{b.hp.} = \frac{PASNE}{33{,}000 \times 4}$$

The S. A. E. formula assumes that all motor car engines would deliver or should deliver their rated power at a piston speed of 1000 ft. per minute, that the mean effective pressure in such engine cylinders would average 90 lb. per square inch, and that the mechanical efficiency would average 75 per cent.

Substituting these values in the above brake horse power equation, and substituting for A its equivalent, $0.7854D^2$, the equation reads:

$$\text{b.hp.} = \frac{90 \times 0.7854D^2 \times 1000 \times N \times 0.75}{33{,}000 \times 4}$$

and combining the numerical values it reduces to:

$$\text{b.hp.} = \frac{D^2 N}{2.489}$$

To make it simpler, the denominator has been changed to 2.5 without materially changing the results.

The formula can be simplified, however, for ordinary use by considering the number of cylinders; thus for the usual four-, six-, and eight-cylinder engines it becomes:

$1.6\, D^2$ = hp. for all four-cylinder motors.
$2.4\, D^2$ = hp. for all six-cylinder motors.
$3.2\, D^2$ = hp. for all eight-cylinder motors.
$4.8\, D^2$ = hp. for all twelve-cylinder motors.

The S. A. E. formula comes very close to the actual horse power delivered by most automobile engines at the piston speed of 1000 ft. per minute. However, at the present time, most of the engines will deliver the maximum power at speeds higher than this, usually around 1500 ft. per minute. As a result, the power which the engines are capable of delivering is greater than that given by the S. A. E. formula. The formula will serve, however, as a means of comparing engines on a uniform basis.

CHAPTER III

POWER-PLANT GROUPS AND TRANSMISSION SYSTEMS

39. Single- and Multi-cylinder Engines.—The first automobile power plant consisted of a one-cylinder engine which gave power impulses at regular intervals of time for the propulsion of the car. Naturally it operated very jerkily and with considerable noise, due to the size of the cylinder and the time between impulses. These facts led to the adoption

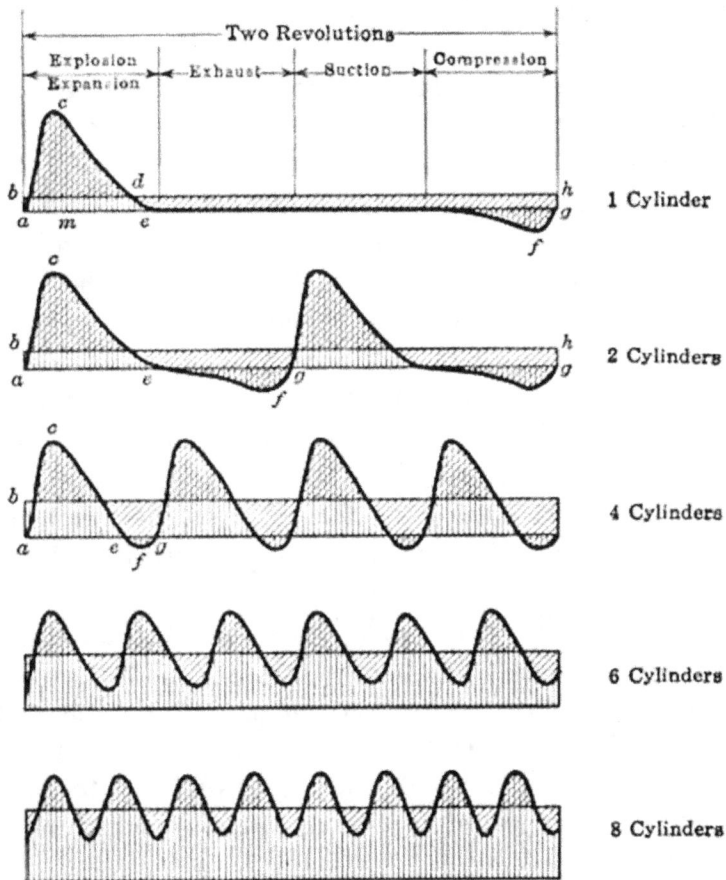

Fig. 54.—Power diagrams.

of the two-, four-, and six-cylinder engines, and quite recently the eight- and twelve-cylinder engines have come into use as automobile power plants.

In Fig. 54 can be seen one of the distinct advantages of the multi-cylinder engine for motor car purposes. The length of the diagram represents two revolutions of the engine crank shaft. The curved line *acefg* represents the variations in the power from a single cylinder. The line *bh* represents uniform power requirement of the car. When the power curve goes above *bh* the engine accelerates and the surplus power is thus stored in the flywheel; when the curve goes below *bh* the flywheel gives up power and the engine slows down.

As the number of cylinders increases, the impulses increase in frequency, the average power is greater, and above four cylinders there is no period during which some cylinder is not delivering power. This means that in a six- or eight-cylinder car, there is no time at which the flywheel must supply all the power required by the car.

The multi-cylinder engine, therefore, furnishes a practically continuous flow of power to the car with little vibration. The increase in the number of cylinders has a tendency to reduce the size of each cylinder and this fact combined with the steady operation of the engine, makes the modern automobile engine a very smooth-running, quiet, power-plant unit.

40. Power Plant and Transmission Arrangements.—Figure 55 shows the arrangement of the Studebaker power plant and transmission system. The engine is placed in the front of the frame, being supported at four points. The clutch, which is of the cone type, is built inside the flywheel, and permits the engine to be disengaged from the transmission system. The propeller shaft, which transmits the power from the engine to rear wheels, is connected to the clutch by means of a universal joint which permits the shaft to receive power and to deliver it to the rear axle.

The change-gear set or transmission is placed on the rear axle just in front of the differential housing which carries the differential gear. The change-gear set permits the relative speed of the engine and car to be changed according to conditions. The chassis diagram indicates the location of the other important parts. Notice the three-quarter elliptic rear springs.

The chassis of the Mitchell "Eight" is shown in Fig. 56. The engine in this case is supported at only three points, one at the front and two at the rear. The clutch is of the cone type operating in connection with the flywheel. It will also be noticed that the change-gear set is placed at the front of the propeller shaft, which then goes directly to the final drive on the rear axle. There is a single universal joint, which is between the clutch and gear set.

The Hollier "Eight" chassis is shown in Fig. 57. Here we see the application of the well-known "unit power plant" in which engine, clutch, and change gears are built into one single unit. This arrangement permits the use of only one universal joint between power plant and rear axle. Notice the cantilever type of rear springs.

In the chassis of the Ford Model T, Fig. 58, use is also made of the

POWER-PLANT GROUPS 45

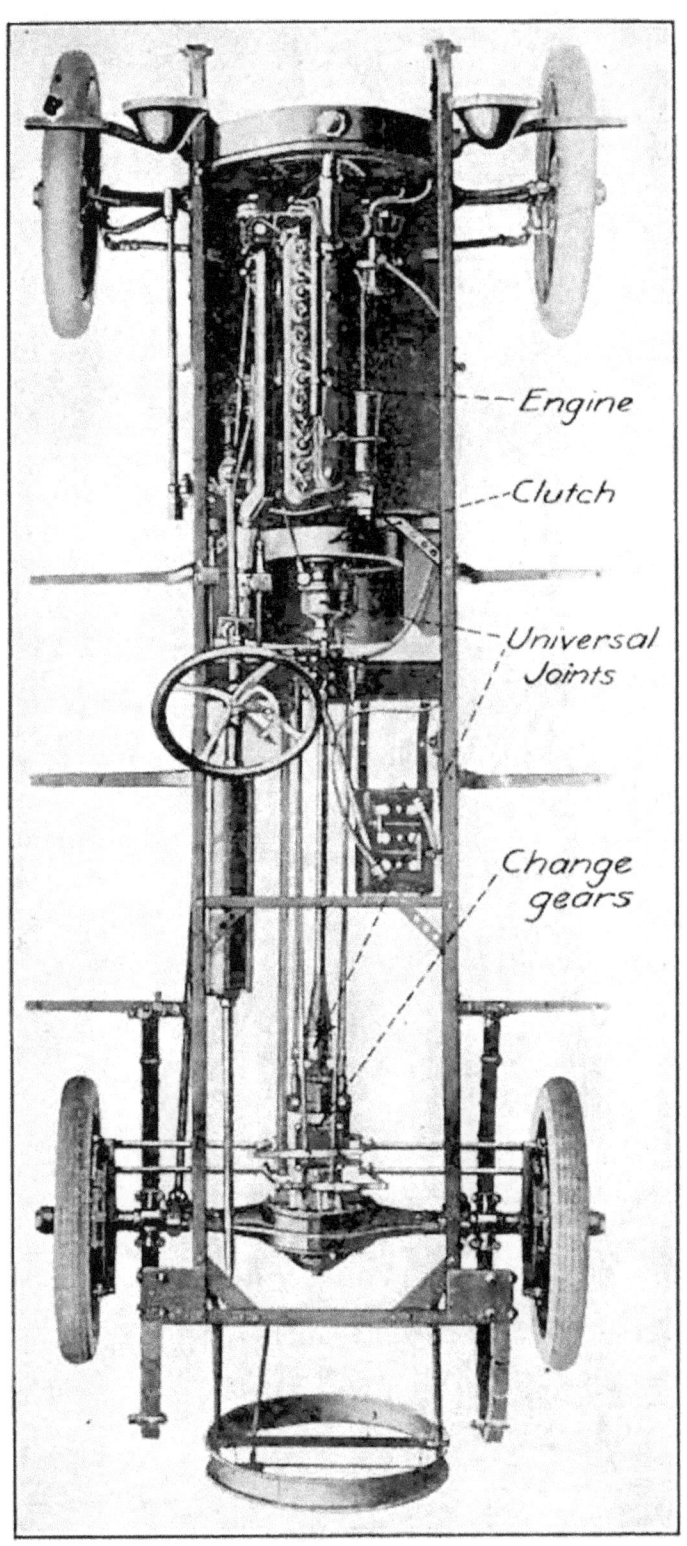

Fig. 55.—Chassis of Studebaker "Six."

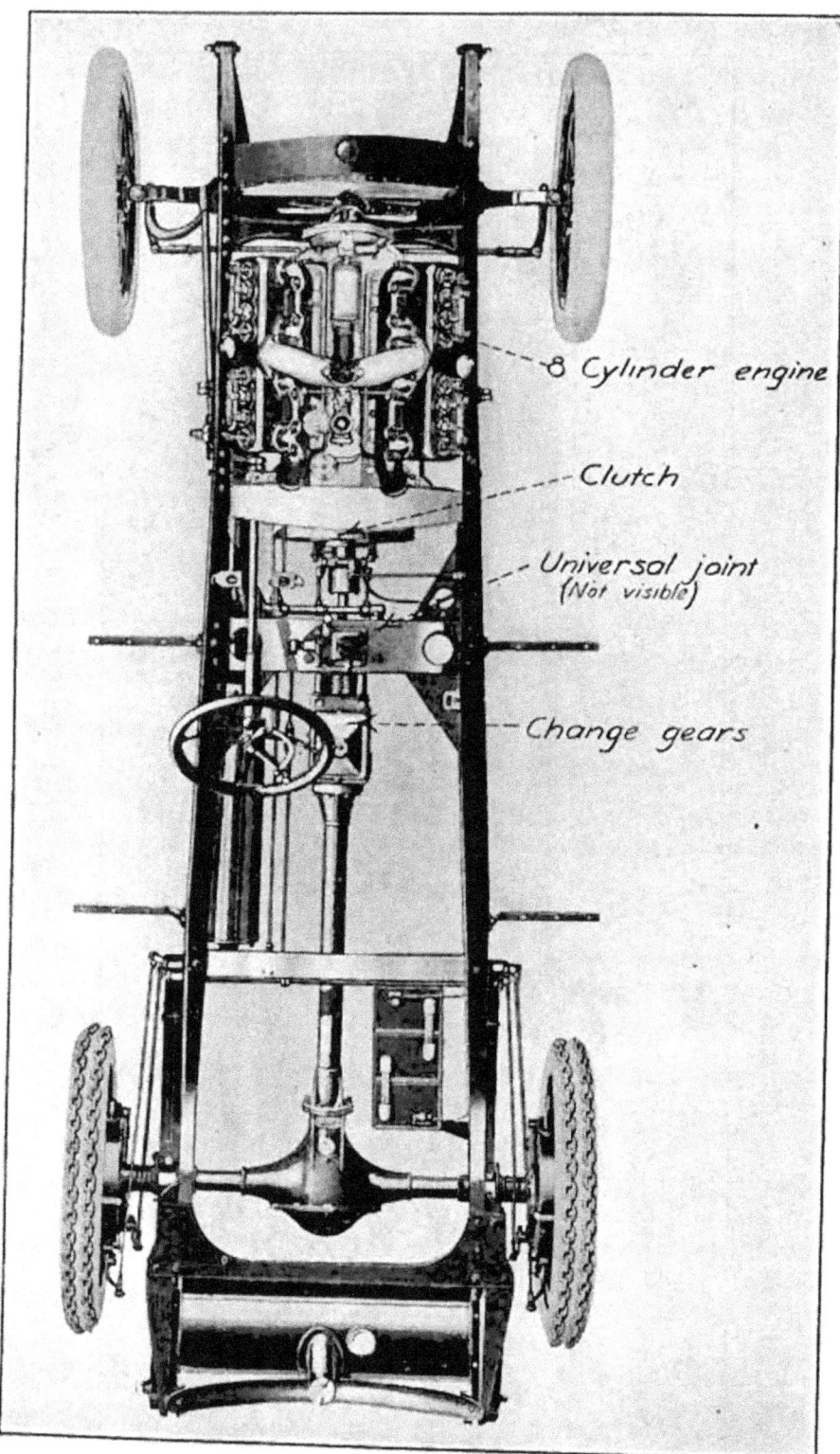

Fig. 56.—Chassis of Mitchell "Eight."

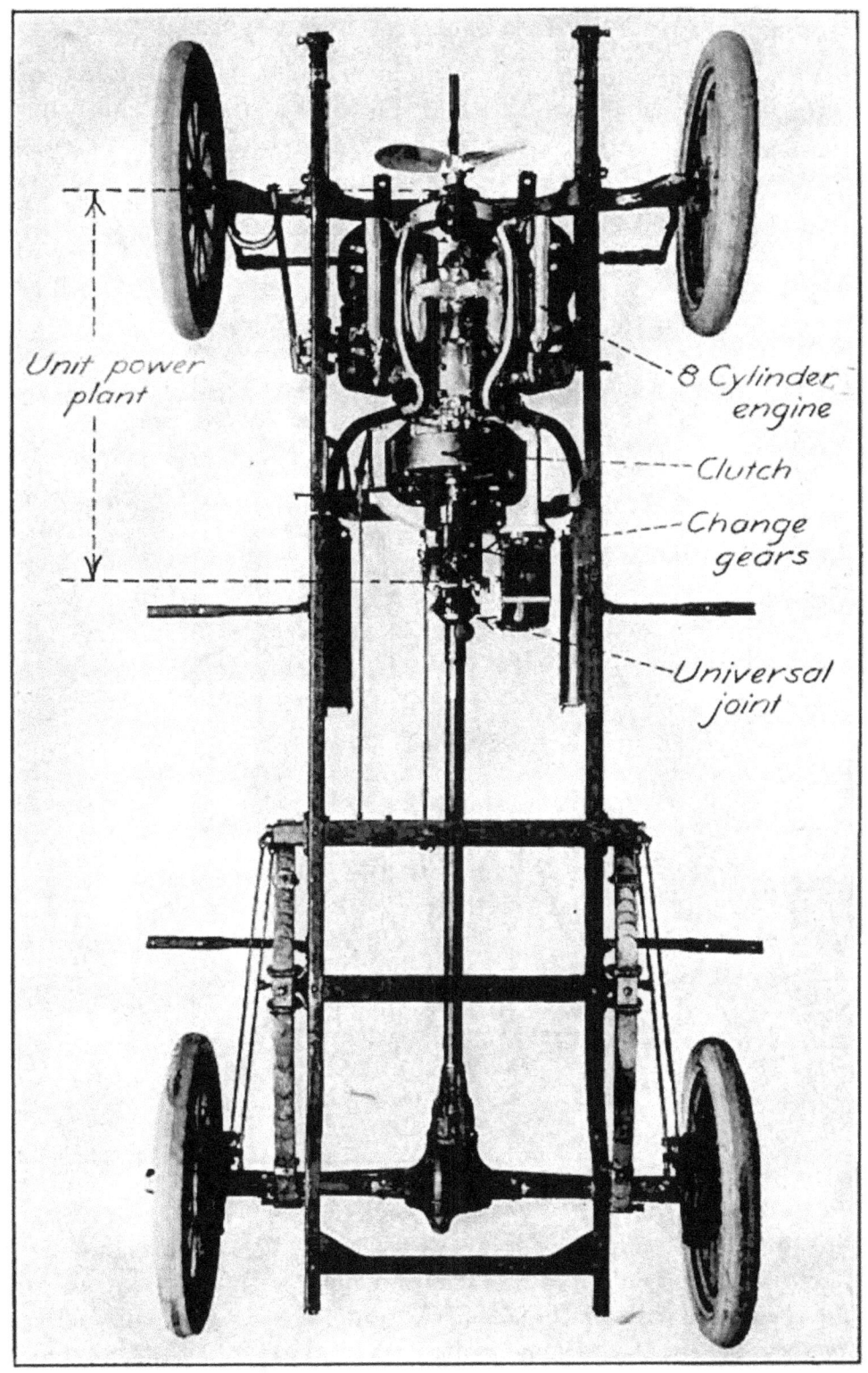

Fig. 57.—Chassis of Hollier "Eight."

"unit power plant" with three-point support. The engine, clutch, and change gears are built together in a single unit and are supported on the frame at only three points. The connection between power plant and rear axle is made by the use of only one universal joint. As will be seen later, this car is equipped with a "planetary" transmission which is built on a principle entirely different from the usual clutches and change-gear sets. The entire rear of the car is supported by an inverted semi-

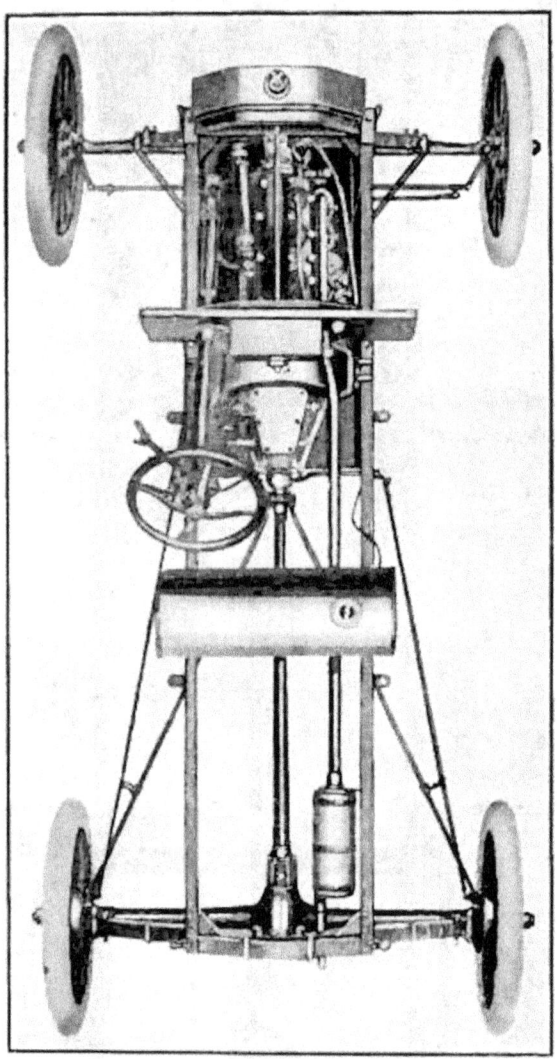

Fig. 58.—Chassis of Ford Model T.

elliptic spring extending over the rear axle. A similar but lighter spring is used in front.

The sectional view of the Lyons-Knight four-cylinder car in Fig. 59 shows very clearly the arrangement of the engine and the transmission groups. The engine is of the Knight type and delivers its power through a plate clutch and through the universal joints and propeller shaft to the

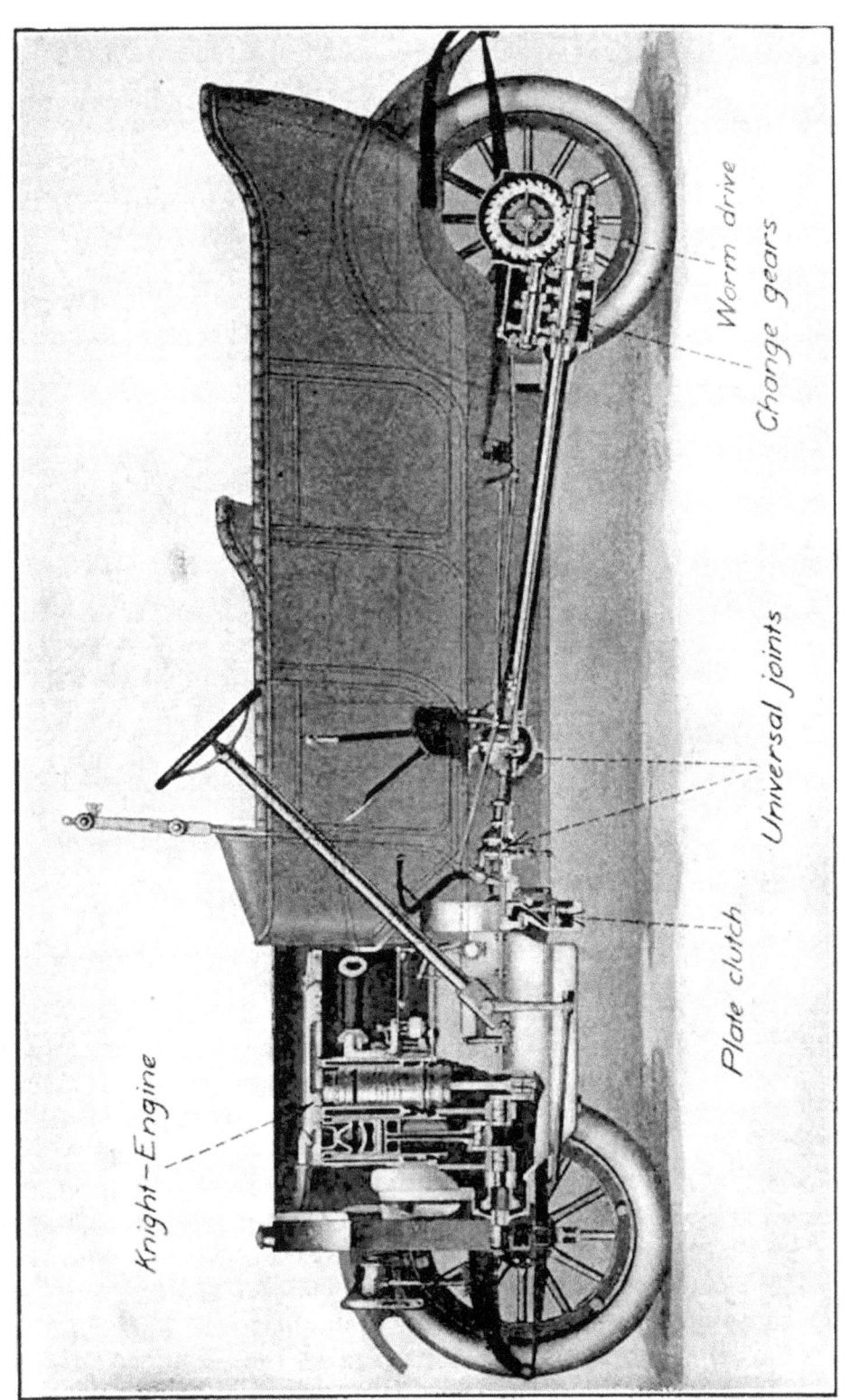

Fig. 59.—Section of Lyons-Knight car.

change gear set built on the rear axle. The final drive from shaft to axle is of the worm type which will be discussed later in the chapter. The clutch control pedal and the change gear control lever are outlined very clearly.

41. Modern Automobile Power Plants.—The automobile power plant includes the engine and all accessories necessary for the production of power. The transmission system includes the mechanism necessary for taking this power furnished by the power plant and transmitting it to the rear wheels.

In most cases, the power plant includes the engine and its component parts such as carburetor, ignition devices, cooling system, etc. and the

Fig. 60.—Four-cylinder Wisconsin engine.

transmission system includes the clutch, change gears, universal joints, differential, and rear axle. When the unit power plant is used, it includes in addition to the engine and its essential component parts, the clutch and the change gears.

Four-cylinder Power Plants.—Figure 60 illustrates a typical four-cylinder automobile engine with the essential parts indicated. The view shown is the exhaust side of the motor, it having the T-head valve arrangement. The cylinders are cast in pairs, two cylinders being in each unit. The water jackets are cast integral with the cylinders. The water connections at the top and bottom of each casting are indicated. The clutch,

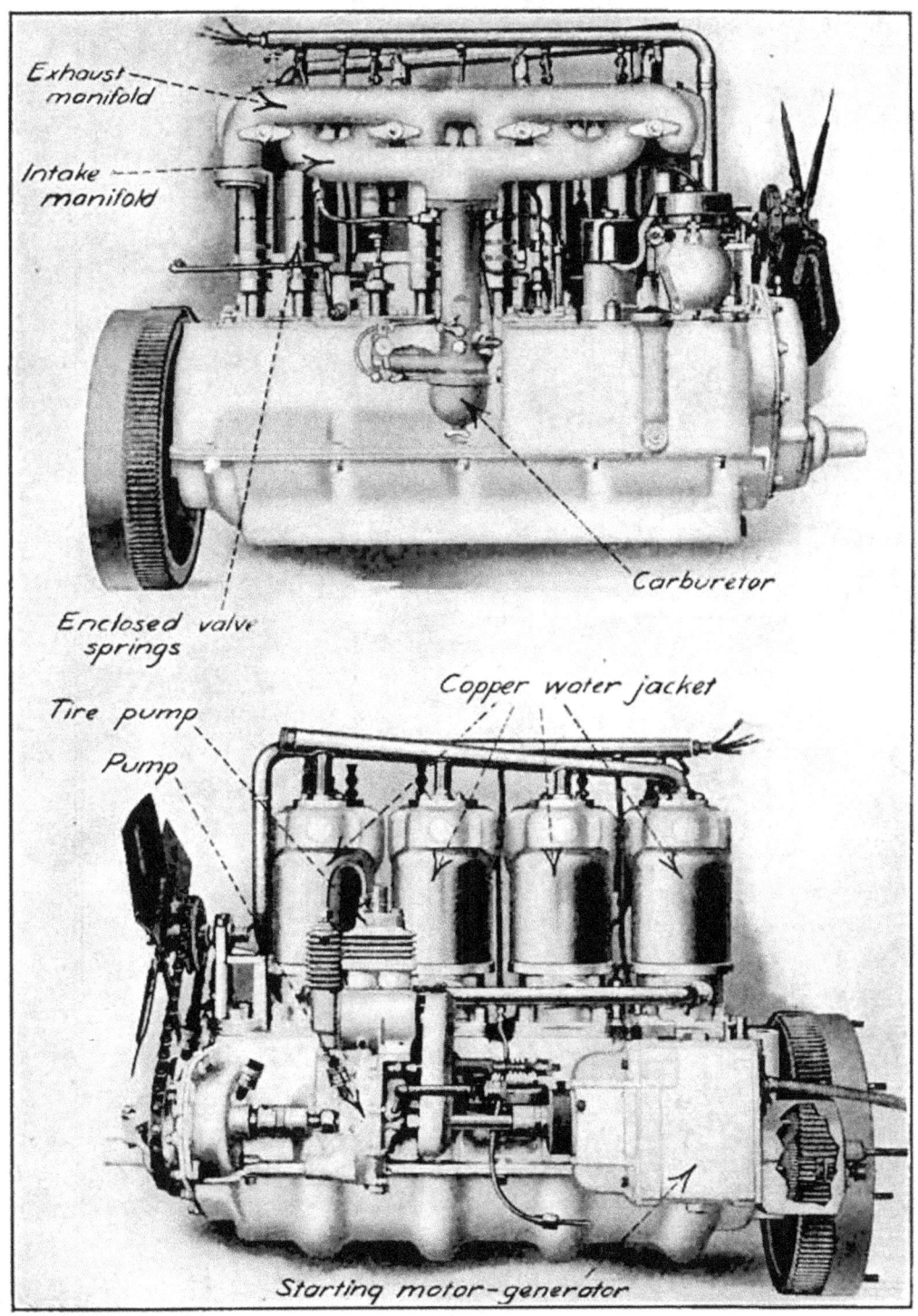

Fig. 61.—The 1914 Cadillac engine.

Fig. 62.—Studebaker "Four" engine.

Fig. 63.—Section of Buda engine.

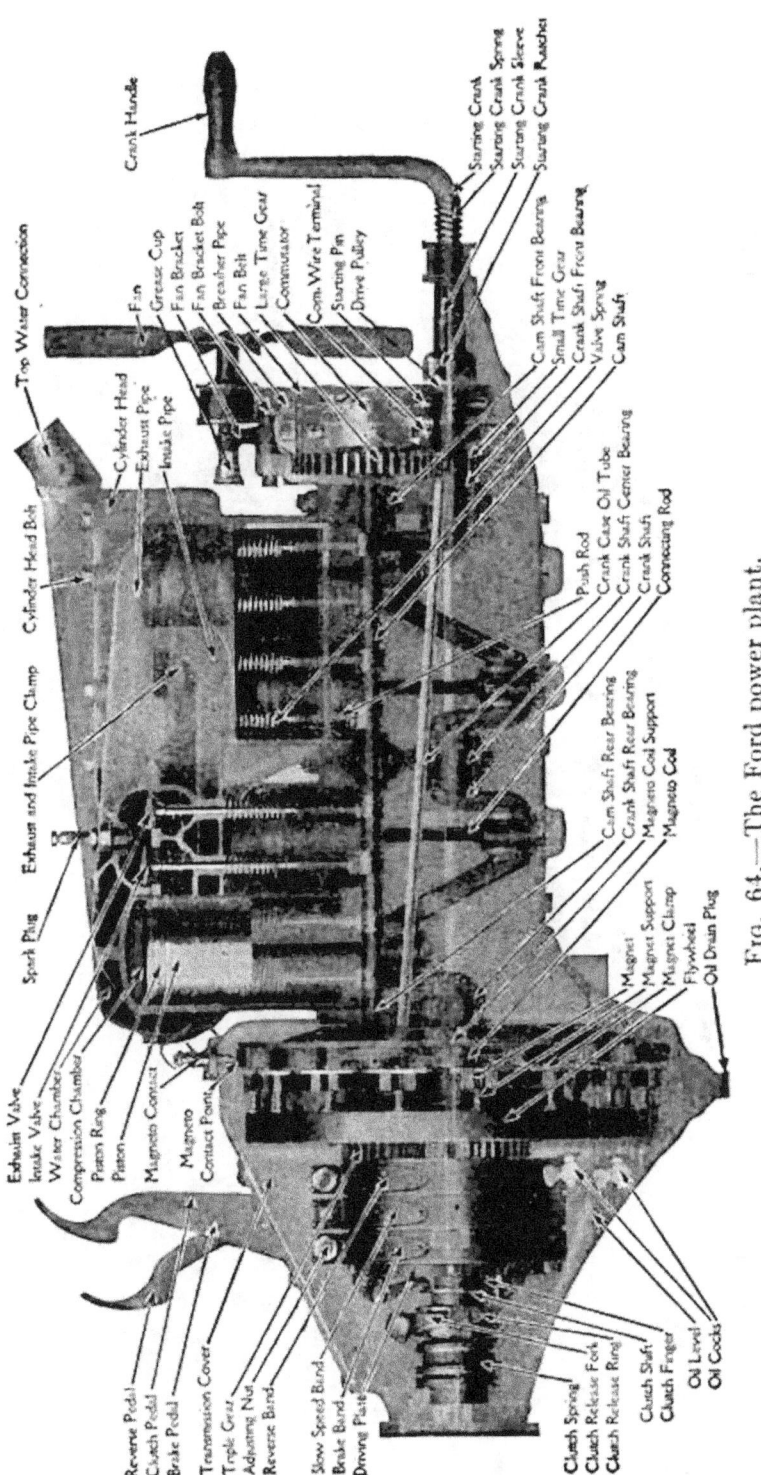

Fig. 64.—The Ford power plant.

one member of which is machined in the engine flywheel, is of the cone type, this being the customary method of applying the cone clutch to the engine.

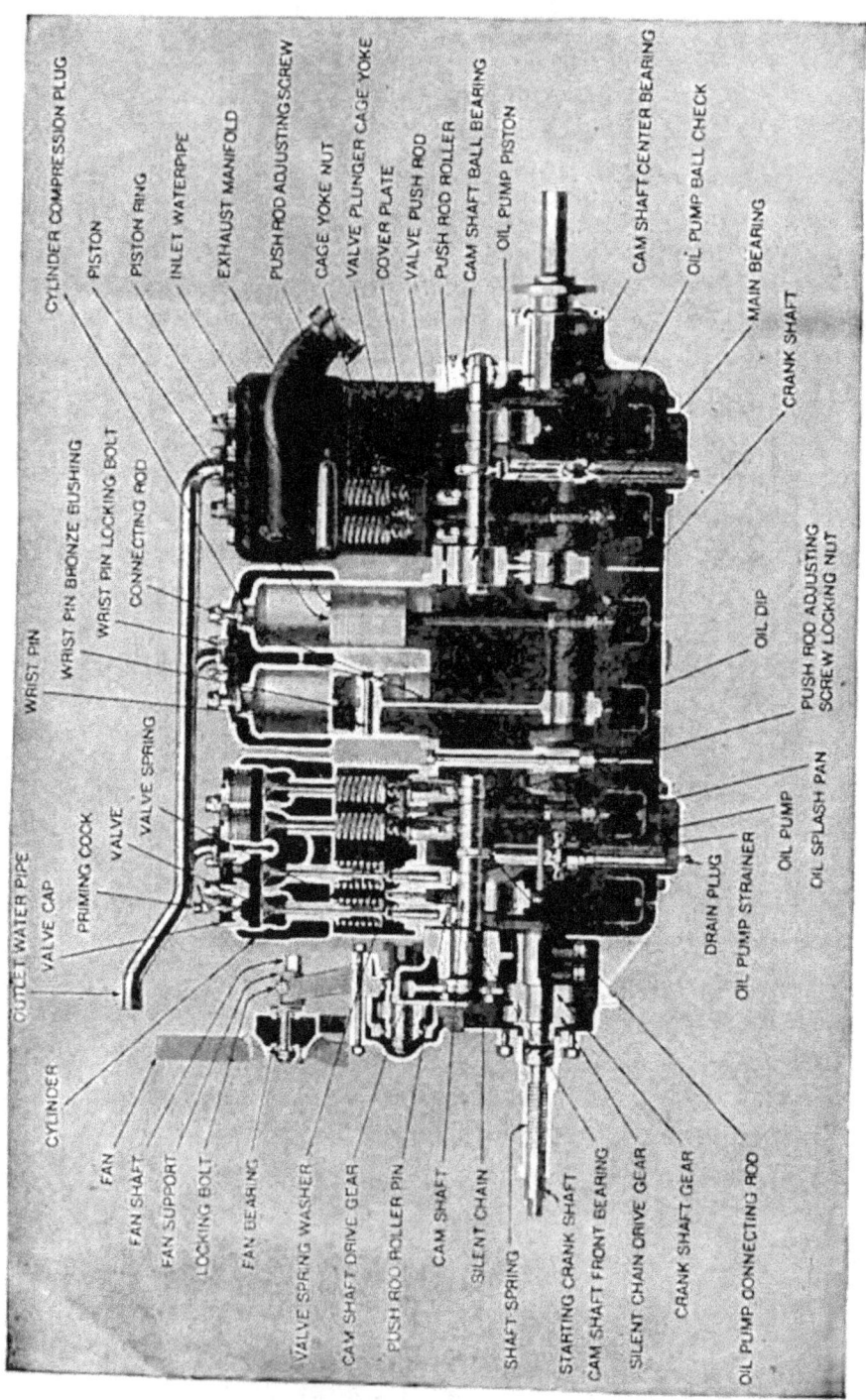

Fig. 65.—Engine of Jeffrey "Chesterfield Six."

The engine of the 1914 Cadillac is illustrated from both sides in Fig. 61. It is of the L-head type, having both intake and exhaust manifolds on the right side. The most prominent feature of this engine is that the cylinders are cast singly with copper water jackets fastened securely around the castings. The single-cylinder castings necessitate a longer engine than if cast in pairs or *en bloc*, but they also make the renewal expense less if a single cylinder is damaged.

Figure 62 is a right-side view of the Studebaker "Four" engine, showing the *en bloc* cylinder construction, in which all cylinders are cast in one piece. This permits the engine to be much shorter than when cast in any other way. The structure is also more rigid, and can be made considerably lighter than when cast singly.

Fig. 66.—Power plant of Mitchell "Six."

The sectional view of a Buda Model T engine in Fig. 63 shows very clearly the internal construction of an engine. This engine is of the L-head type with only one cam shaft. The crank shaft is of the conventional three-bearing type, *i.e.*, with a bearing at each end and one at the center.

The Ford unit power plant is shown in section in Fig. 64 with all parts fully designated. The magneto, change gears and clutching arrangement are of considerable interest and will be discussed under the proper headings. As will be remembered, this power plant has three-point support.

Six-cylinder Power Plants.—The Jeffrey six-cylinder power plant is shown in section in Fig. 65. The cylinders are cast in pairs, thus permitting the use of a four-bearing crank shaft. In the pair of cylinders at the left, the section is taken through the valves so as to show the cams, push rods, springs, and valves. The center pair is sectioned through the center of the cylinders so as to show the pistons, pins, and connecting rods. The valve arrangement is of the L-head type.

The engine of the "Mitchell Six of '16," Fig. 66, has the six cylinders cast "*en bloc,*" which gives a very compact and rigid construction of pleasing appearance. The cylinder head can be removed in one piece for the purpose of cylinder and valve examination.

The Franklin motor, Fig. 67, represents a very interesting and unique design, having overhead valves and air-cooling. The cylinders are cast singly and each is air cooled by a system of cast ribs and air cooling, doing away with the water jackets around the cylinders. The

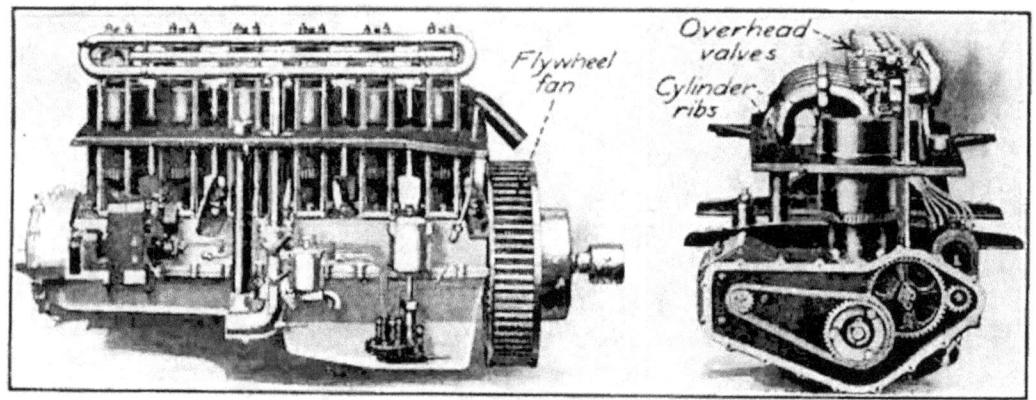

Fig. 67.—The Franklin air-cooled engine.

air is drawn downward around the cylinder ribs by the suction of the flywheel fan.

42. Constructional Features of Four- and Six-cylinder Engines.—The essential differences of construction in the various four- and six-cylinder engines, outside of the methods of cylinder construction and valve arrangement, consist in the construction and arrangement of the cam and crank shafts. Figure 68 is a conventional four-cylinder crank shaft, shown with connecting rods and pistons attached. There are three main bearings, as indicated. The connecting rod bearings are all in the same plane, bearings Nos. 1 and 4 being just 180° from Nos. 2 and 3. This means that the Nos. 1 and 4 pistons are in the same position in the cylinders at the same time. Likewise Nos. 2 and 3 are in the same position. If No. 1 piston is on the compression stroke, No. 4 must necessarily be on the exhaust stroke and Nos. 2 and 3 on the suction and explosion strokes.

The order of firing in a four-cylinder engine must be in the order 1, 3, 4, 2 or 1, 2, 4, 3.

The five-bearing crank shaft for a four-cylinder engine has main bearings between all the cranks. Figure 69 shows the five-bearing crank shaft

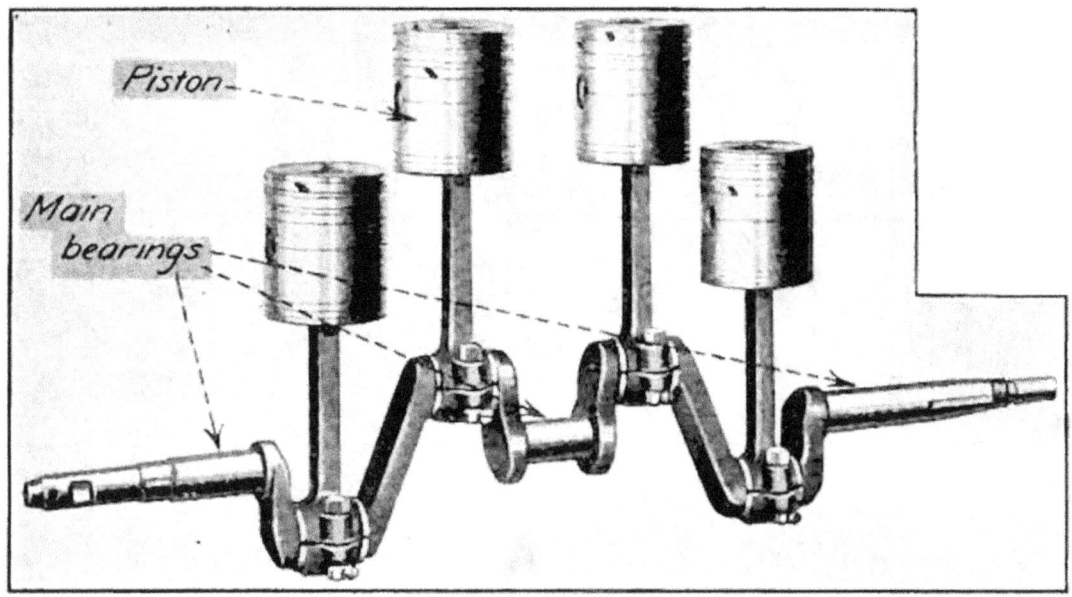

Fig. 68.—Three-bearing, four-cylinder crank shaft.

in place on the 1914 Cadillac four-cylinder engine. This type of crankshaft construction is especially adapted to an engine with individually cast cylinders.

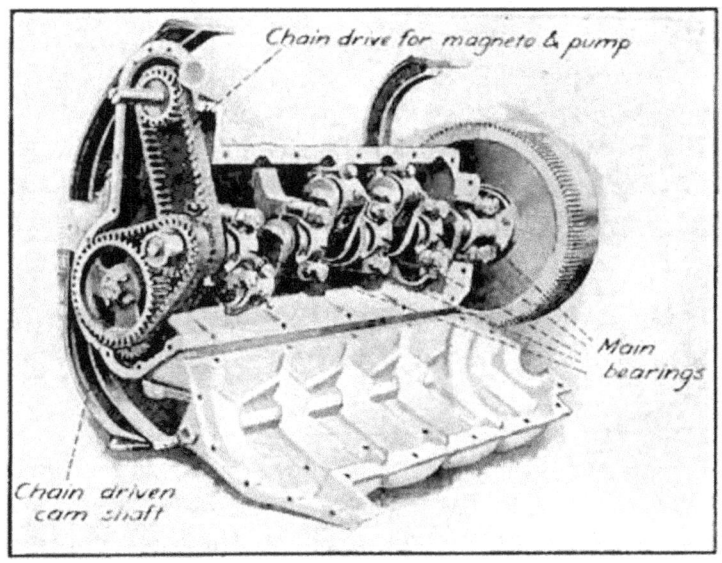

Fig. 69.—Five-bearing, four-cylinder crank shaft in position.

The crank shaft for a six-cylinder engine is arranged as shown in Fig. 70. Cranks 1 and 6, 2 and 5, 3 and 4 are in pairs and are spaced 120°

apart. The pistons in the paired cylinders are always in the same relative positions in the cylinders. The firing order of the cylinders is usually 1, 5, 3, 6, 2, 4 or 1, 2, 3, 6, 5, 4. This crank has four main bearings. The

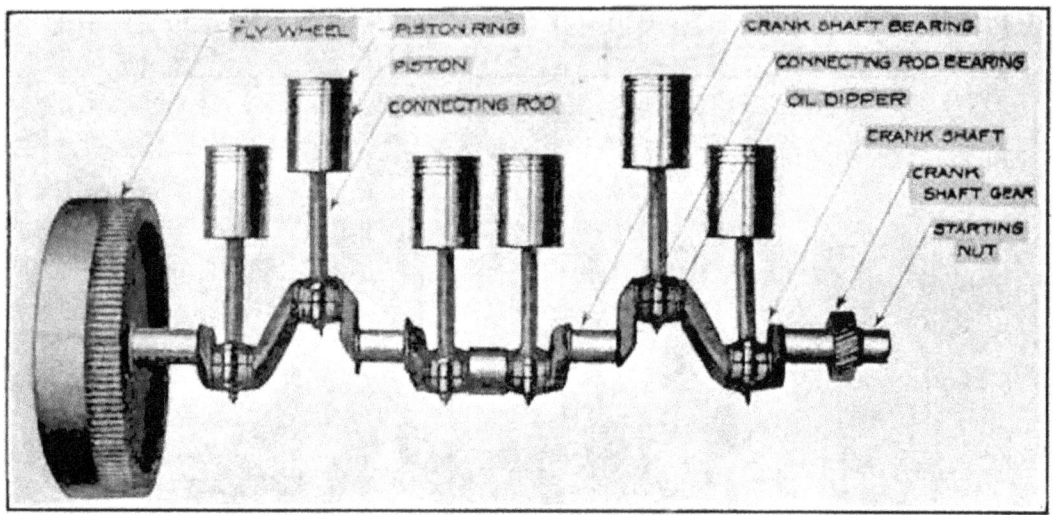

Fig. 70.—Four-bearing, six-cylinder crank shaft.

shaft shown in Fig. 71 has only three main bearings. The arrangement of the cranks is the same as in the previous case.

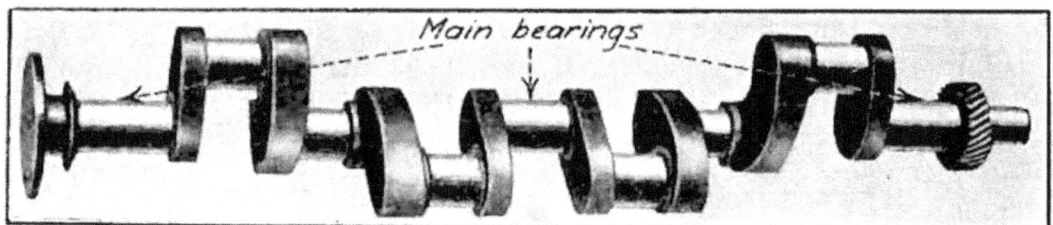

Fig. 71.—Three-bearing, six-cylinder crank shaft.

In Figs. 72 and 73 are illustrated the two general methods of cam shaft construction. Figure 72 is a one-piece cam shaft, the cams and shaft

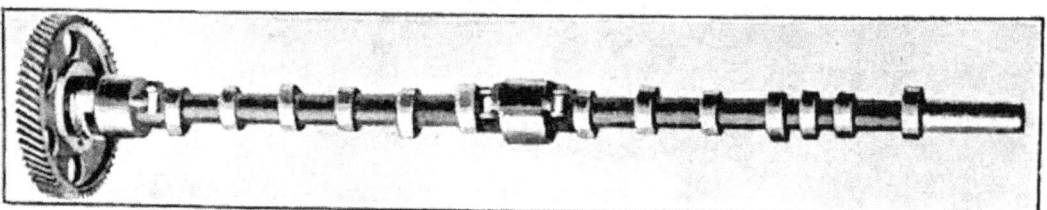

Fig. 72.—One-piece cam shaft.

being made of one solid bar of steel. This is the more common method of construction. The assembled cam shaft, Fig. 73, on which the individual cams are pinned or keyed is used at present in very few cases. The objection to this type of shaft is that the cams may become loose on the shaft and give considerable trouble. For an L-head engine, a single cam shaft on one side of the engine carries both inlet and exhaust cams. For

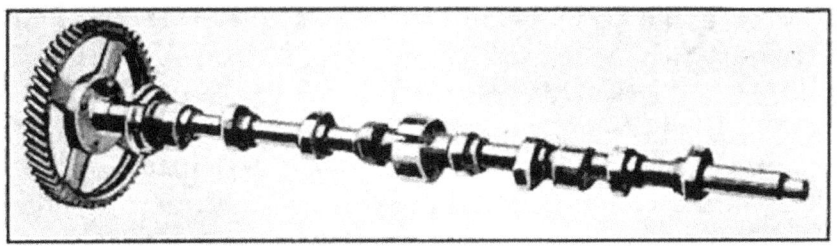

Fig. 73.—Assembled cam shaft.

Fig. 74.—Cadillac eight-cylinder V-type engine.

a T-head engine, however, one cam shaft carries the inlet cams on one side of the engine and another shaft carries the exhaust cams on the other side

The cam shafts are driven at one-half crank shaft speed. The drive can either be by a silent chain, such as shown for the 1914 Cadillac in Fig. 69, by spur gears such as in the Ford Model T shown in Fig. 64, or by helical gears such as shown in Figs. 72 and 73.

43. Eight- and Twelve-cylinder Power Plants.—In the four-cylinder engine, there is a power stroke every one-half revolution, but during a small interval at the end of each power stroke no power is being delivered by the engine. This means short periods in the operation of the engine in which the flywheel must supply all the power. In the six-cylinder engine,

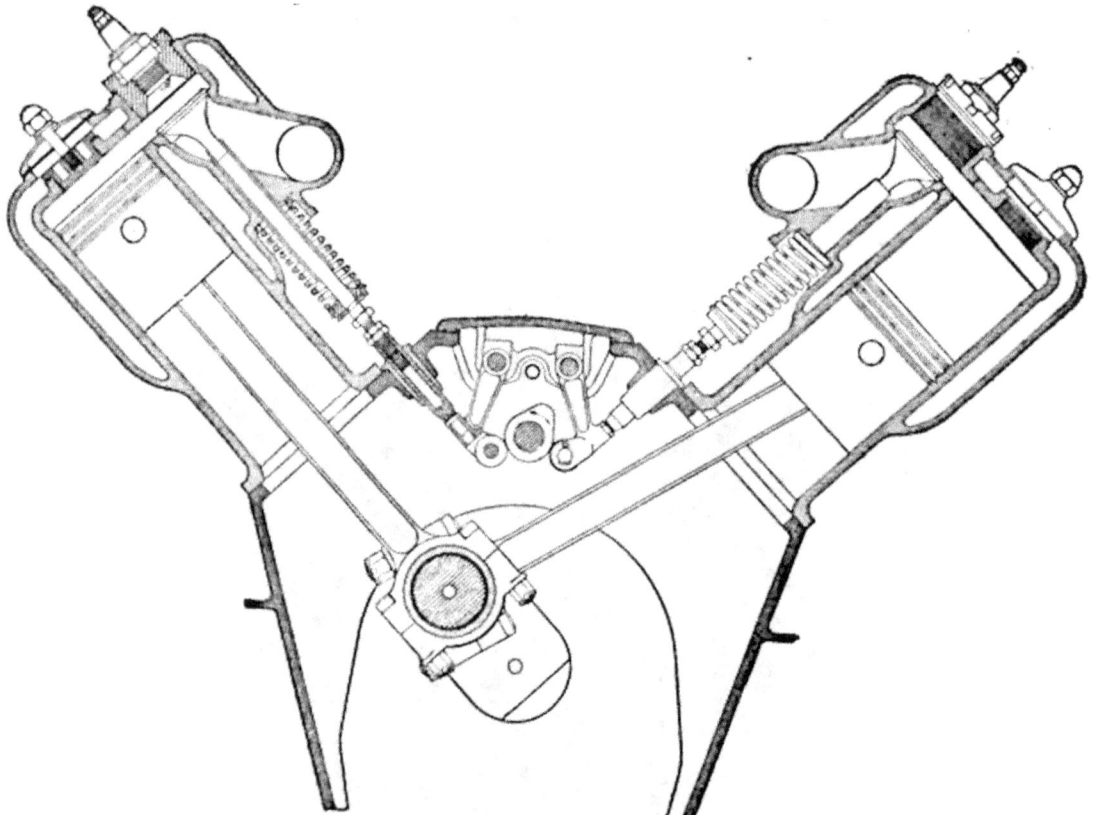

Fig. 75.—Sectional view of Cadillac eight-cylinder engine.

there is a power stroke every one-third revolution and, as a result, there is an overlapping and a more continuous flow of power (see Fig. 54). The impulses come oftener and, consequently, reduce the vibration. The same effect is carried further in the eight-cylinder engine which gives a power stroke every one-fourth revolution. The parts are considerably lighter and this aids in reducing the vibration. Most of the eight-cylinder engines are built in the V-type and this method of construction adds to the smoothness of operation.

Cadillac Eight-cylinder Engine.—Figure 74 is a front-end view of the Cadillac eight-cylinder engine. The cylinders are arranged in blocks of

POWER-PLANT GROUPS

four each, placed in a V-shape at an angle of 90°. A cross section of two opposite cylinders is shown in Fig. 75. The engine is of the L-head type with the valves on the inside of the V. One cam shaft placed directly above the crank shaft operates all of the sixteen valves by means of the rockers as shown. Eight cams serve to operate the sixteen valves, as

Fig. 76.—A pair of Cadillac connecting rods.

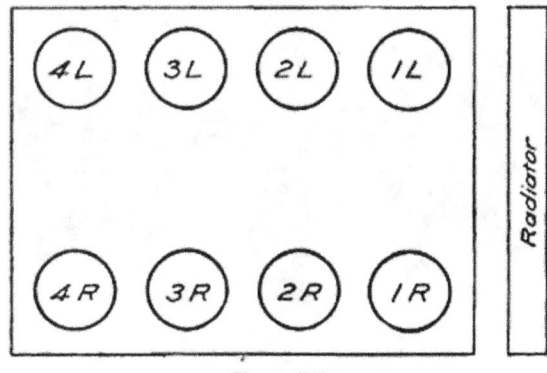

Fig. 77.

one cam operates a valve in each group. The cam shaft is carried by five bearings and has a silent chain drive as shown in Fig. 74.

The crank shaft is like a conventional four-cylinder shaft with three main bearings. There are only four crank pins, two connecting rods, one from each group, bearing on the same crank. One of the rods, Fig. 76, is forked, while the other is perfectly straight, fitting in between the fork. The split bearing shown at the right fits directly over the pin. The forked rod fits over this bearing and is pinned to it, so that the rod and bearing work together. The other rod fits in the center surface of the bearing and

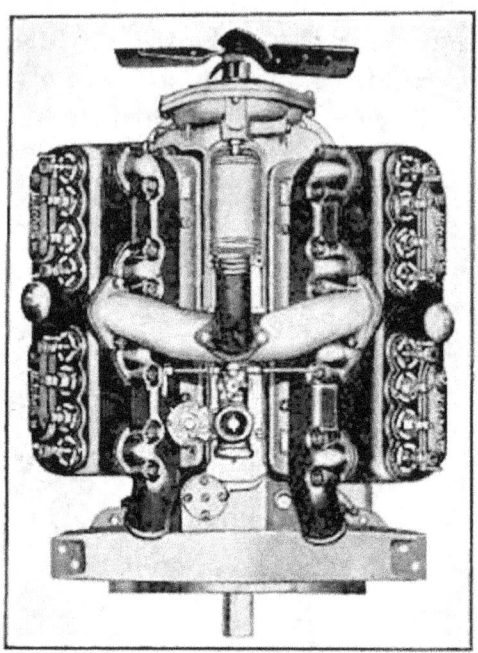

Fig. 78.—Top view of Mitchell "Eight" engine.

Fig. 79.—Front view of Mitchell "Eight" engine.

runs on it. The arrangements permit the length of the crank shaft to be no greater than in a four-cylinder engine.

The order of firing of the eight cylinders alternates from one side to the other. If the cylinders be numbered as shown in Fig. 77 the firing order is as follows: 1-L, 2-R, 3-L, 1-R, 4-L, 3-R, 2-L, and 4-R. The horse power rating of the Cadillac Eight is 31.25 according to the S. A. E. formula. On dynamometer test, however, it has developed 70 hp. at a speed of 2400 r.p.m.

Mitchell Eight.—The Mitchell Eight is constructed on the same general principle as the type previously mentioned. The cylinder groups are placed in a V of 90°. The valves are placed on the inside of the V and

Fig. 80.—Engine of Packard "Twin Six."

are operated by means of eight cams on a single cam shaft mounted above the crank shaft. The cylinders are slightly staggered and two connecting rods are mounted side by side on each crank instead of using the forked construction.

The engine is rated at 48 hp. The cylinders are 3-in. bore by 5⅛-in. stroke. The top and front-end views are shown in Figs. 78 and 79.

The Packard Twelve-cylinder Engine.—The twelve-cylinder unit power plant of the Packard car is shown in Fig. 80. The twelve cylinders are cast in two blocks of six, arranged in V-type with an included angle of 60°. The cylinders have a 3-in. bore and a 5-in. stroke with L-head valve arrangement. The left block of cylinders is set forward of the right set by

1¼ in. in order to permit the lower end of the connecting rods of opposite cylinders to be placed side by side on the same crank pin. In addition, this arrangement permits the use of a separate cam for each valve, making 24 cams on the cam shaft. The single cam shaft is placed directly above the crank shaft. The crank shaft is of the usual six-cylinder type supported by three main bearings.

Advantages Claimed for Eight- and Twelve-cylinder Motors.—The chief advantages claimed by the eight- and twelve-cylinder motors are smooth running, lack of vibration, rapidity of pick-up, and wide range of activity

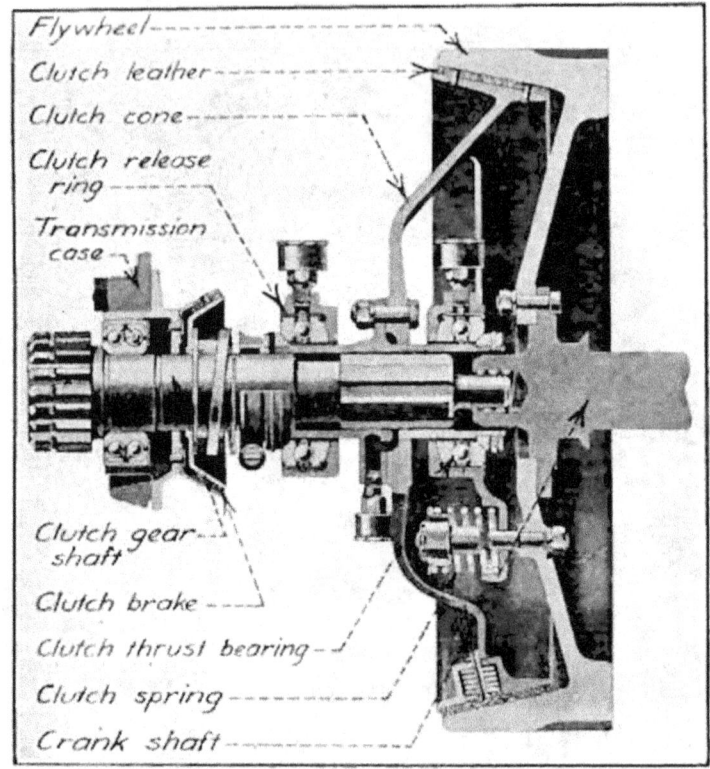

Fig. 81.—Buick cone clutch.

on high gear. It is possible with either of these types to run almost entirely on high speed under all conditions.

44. Clutches.—The gasoline engine must be set in motion before it will take up its cycle and generate power. This fact prevents it from being started under load and, consequently, means must be provided for detaching the engine from the rest of the mechanism for starting before the load is thrown on. This mechanism for detaching the engine from the remaining part of the power and transmission system is called the "clutch." There are in use at the present time two general types of clutches, the cone type and the disc type.

The Cone Clutch.—Figure 81 illustrates the cone clutch as used in the Buick car. It consists of a leather-faced aluminum cone which is held tightly against the inside of the tapered rim of the flywheel by four springs carried on a spider. The aluminum cone is mounted on a steel sleeve which can slide back and forth on the clutch gear shaft to disengage or engage the cone with the flywheel. A grooved ring at the rear end of the sleeve connects the clutch to the clutch pedal. A small brake, attached to the transmission case, serves to keep the clutch from spinning after it is released. Four small spring plungers, located under the leather, force it out at these points and prevent grabbing when the clutch is let in.

In operation, pressure on the clutch pedal is transmitted by a connecting link and clutch release shaft to the yoke operating on the ball-bearing release ring, which pulls the clutch back out of engagement with the flywheel. The small brake now holds the clutch stationary, while the clutch spider and springs continue to turn with the flywheel until the clutch is again engaged. When in full engagement, the clutch and flywheel turn as a unit, transmitting the power through the gear set to the rear axle.

Multiple Disc Clutches.—The multiple disc clutch is built in two types—the dry plate and wet plate. Figure 82 is a sectional view of the dry plate type of clutch as used on the Hudson. It consists of a series of alternate driving and driven discs. The driving discs receive their power from the flywheel by four studs, one of which shows in the cut. These discs are steel stampings.

The driven discs are also steel stampings but are somewhat thicker and have holes into which cork inserts are pressed. The driven discs drive the inner drum by means of a series of grooves or slots.

The driven and driving discs are pressed together by the clutch spring shown. When it is desired to release the clutch, the foot pedal compresses the clutch spring and the plates separate, permitting the driving members to run independently of the driven members. As in all clutches, the power is transmitted entirely through a frictional contact. The cork inserts are used because they are soft and at the same time have a great adhesive property, even if they become soaked with oil. The advantage of this type of clutching arrangement is that a large frictional surface can be obtained with a comparatively small clutch diameter. In the cone type this diameter must necessarily be large in order to get the necessary friction surface on the one surface in contact. In letting in the plain cone type of clutch, there is also the possibility of a more sudden engagement than with the multiple disc type. This has been overcome by the use of the springs under the leather, as shown in Fig. 81.

The wet plate clutch is constructed on the same general principles as the dry plate clutch, the essential difference being that it runs in a bath of oil. When the clutch is released, an oil film covers the entire surface of the plates and, when the clutch is thrown in, this film of oil is gradually squeezed out, permitting a very easy and gradual engagement. In the winter time, the oil may be unusually heavy and this prevents a quick engagement. This can be overcome by thinning the clutch oil with kerosene.

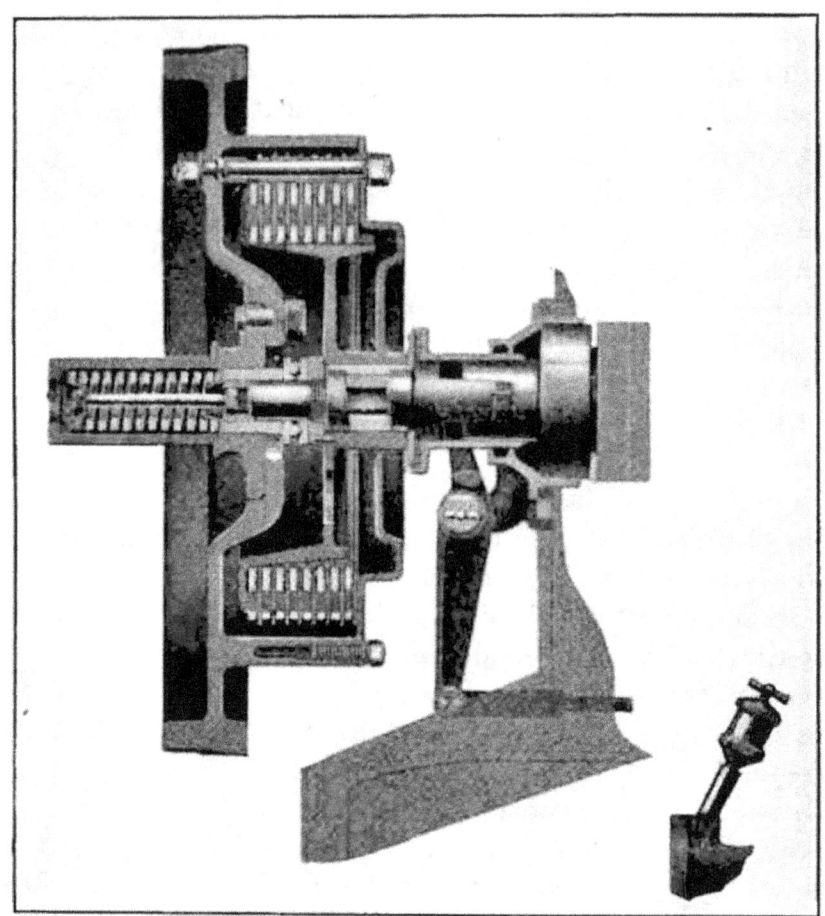

Fig. 82.—Hudson dry plate clutch.

45. Change Gear Sets.—The change gear set is for the purpose of permitting different speed ratios between the engine and the car. When starting, the engine must run comparatively fast and the car slow. When the car gets under way, the relative speed of car and engine must be changed in order to get efficient operation.

Figure 83 is the gear set used on the Jeffrey car. The right shaft is driven by the clutch; attached to this shaft is the drive gear which at all time drives the lay-shaft drive gear fastened to the lay-shaft. The lay shaft in addition carries four fixed gears as shown. The main drive shaft has one end bearing rotating within the main drive gear. Consequently the drive gear and main shaft can run independently of each other. The main shaft carries two sets of sliding gears, the names and purposes of which are indicated. These two sets are operated by two shifter yokes which lead to the gear control lever in the car. This gear set provides four forward speeds and a reverse speed. This type is known as the "selective sliding gear set," because, as the name indicates, any one of the speeds can be selected at will, in contrast to the "progressive sliding gear set" in which the speeds must be taken in succession.

Figure 84 illustrates the gear positions for the various speeds obtained in the Studebaker three-speed-and-reverse gear set. The white arrows indicate the gears through which the power is transmitted for the different speeds.

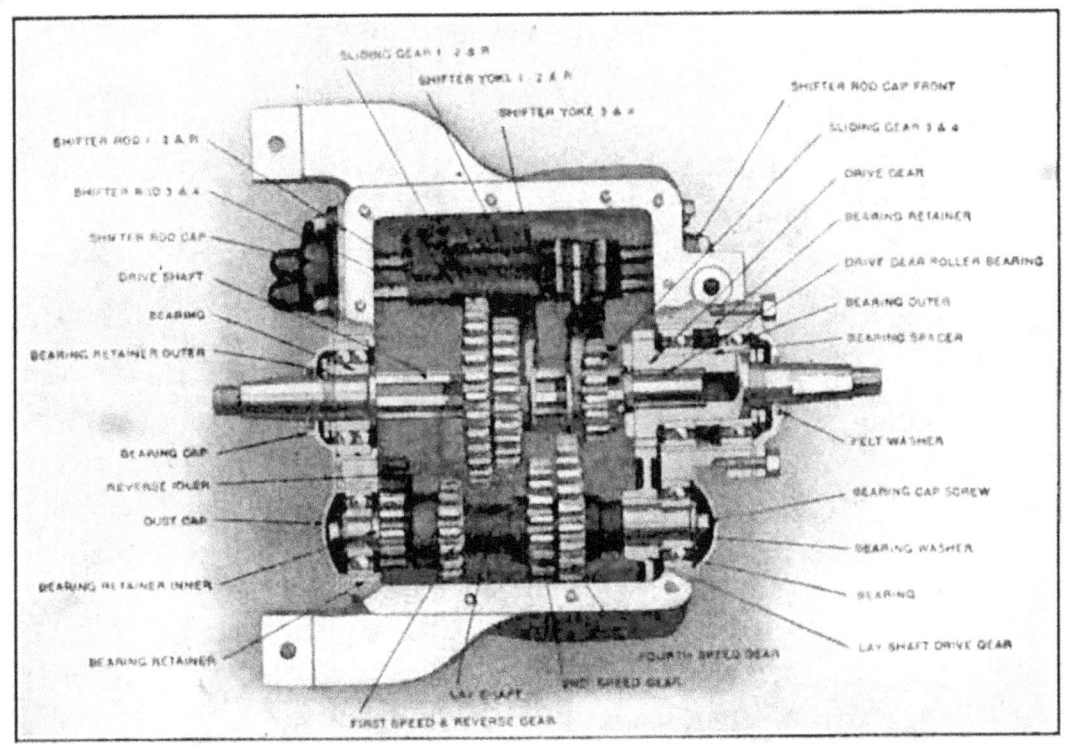

FIG. 83.—Jeffrey gear set.

46. Planetary Gearing.—This type of combined clutch and change gears, such as used on the Ford Model T, is especially adapted to light cars in which two forward speeds are sufficient. The gears are not shifted into or out of mesh for the different speeds, as in the sliding gear set, but they are always in mesh, as shown in Fig. 85. On high gear, the entire mechanism is clamped solidly together by the clutch and revolves as a single mass with the flywheel. The clutch is of the multiple disc type, running in oil. The flywheel has three studs, each of which carries three gears of different sizes fastened together to form what is called a "triple gear." These triple gears mesh with three gears of different sizes in line with the engine shaft. The inner one, next to the flywheel face, is fast-

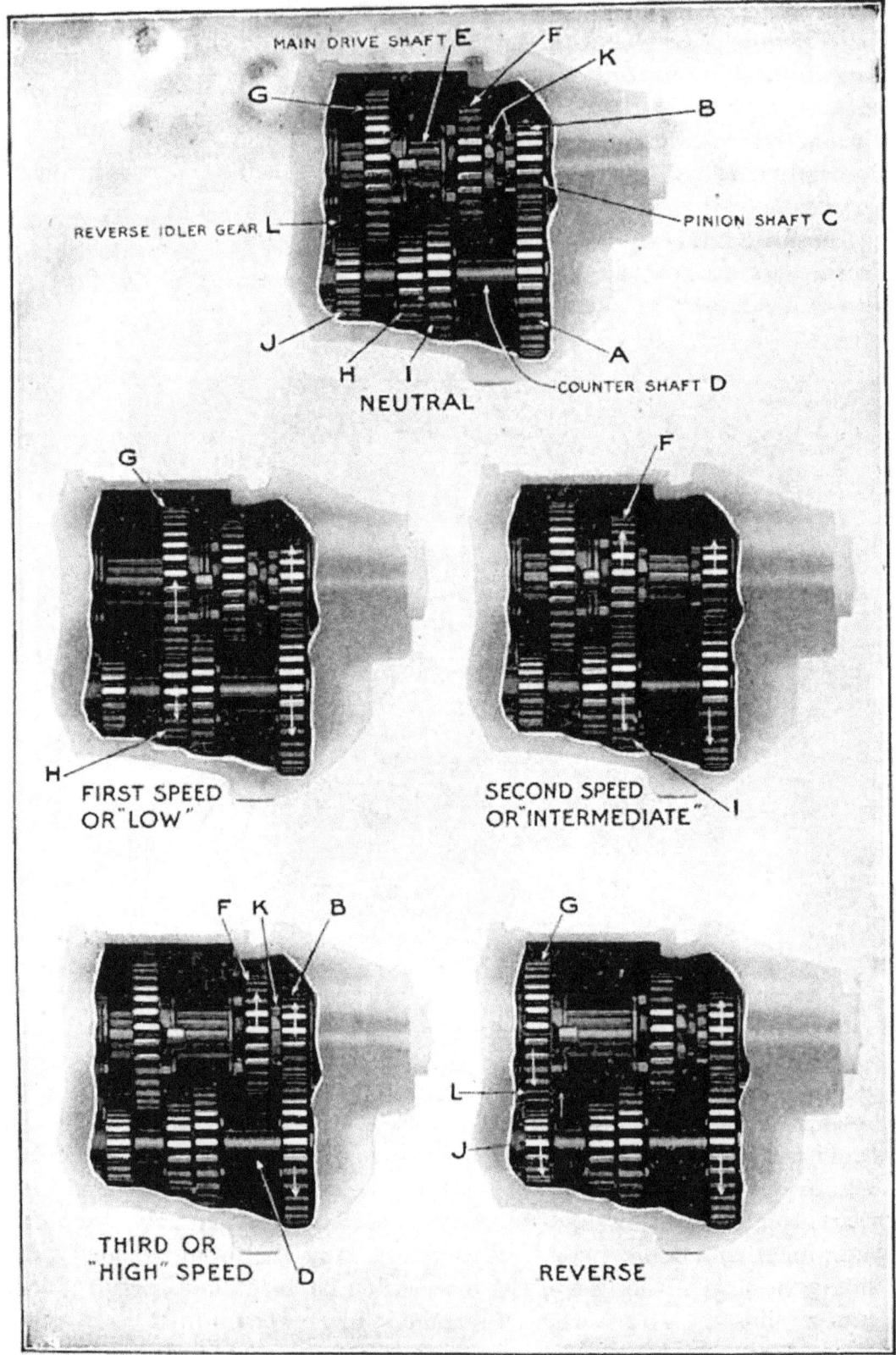

Fig. 84.—Positions of gears in Studebaker three-speed-and-reverse gear set.

ened to the drive shaft which delivers the power through to the rear axle. The other two central gears float on the drive shaft and are connected to the two drums nearest to the engine. Surrounding these drums, but not shown in the figure, are brake bands which can be tightened by foot pedals. These can be seen in Fig. 64. If the slow-speed drum is gripped, the second of the three central gears will be held stationary. This makes the triple gears rotate on their studs as the flywheel revolves. In doing this, they drive the inner central gear, or the driving gear, slowly forward,

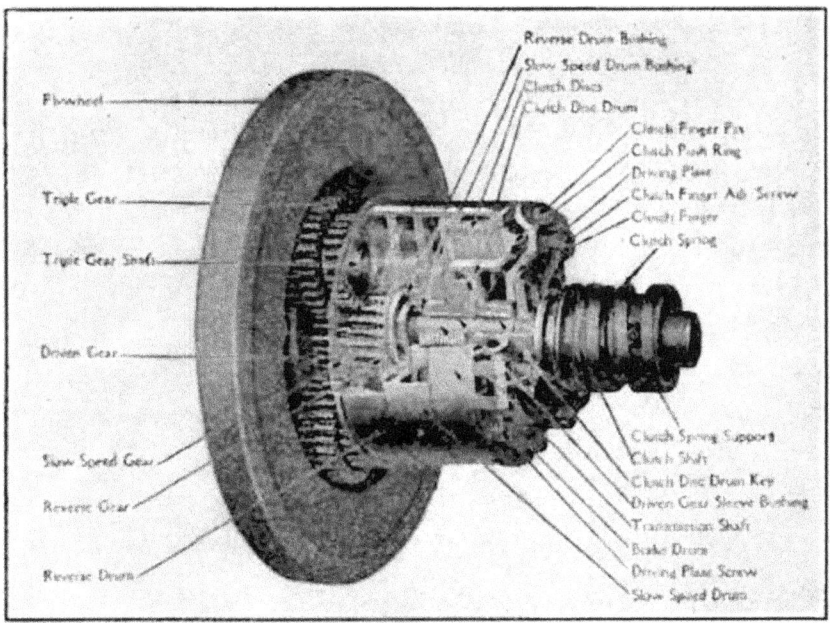

Fig. 85.—Ford planetary transmission.

due to the differences in the sizes of the gears. If the middle drum is gripped instead, by pushing on the reverse pedal, the larger of the central gears is held. This makes the triple gears revolve again on their studs as the flywheel revolves, but since this reverse gear is larger than the drive gear, the motion of these triple gears will turn the drive gear slowly backward. For high speed, the entire mechanism is gripped solidly together so that it revolves at engine speed. The third drum is used for a service brake.

47. Universal Joints and Drive Shaft.—The use of one or more universal joints between the power plant and the rear axle is necessary, as can be seen in Fig. 59, in order to provide for the lower position of the rear axle and also to allow for the spring action between the axle and the frame which carries the power plant. The universal joint permits this to be done with very little loss of power. Figure 86 shows the propeller shaft or drive shaft of the Jeffrey car with its universal joints. A square block in the center of the universal joint fits between the jaws of two forks, one of which is connected to the power plant and the other is attached to the end of the drive shaft. The flexible connection of these forks to the block permits the drive shaft to oscillate freely with the rear axle and yet continue to receive and transmit power.

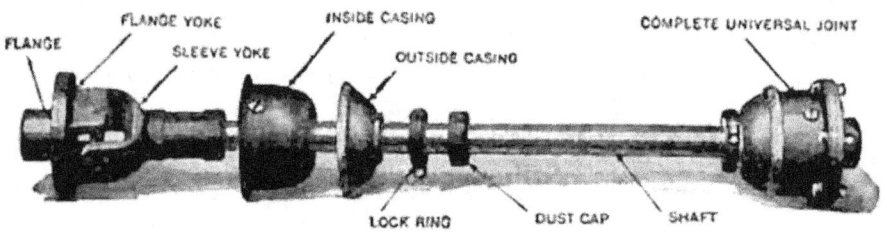

Fig. 86.—Jeffrey propeller shaft and universal joints.

48. Final Drive.—The final drive to the rear axle is accomplished by means of bevel, spiral-bevel, or worm gearing. The direction of the power transmission must be changed through a right angle at this point. Figure 87 shows the bevel gear final drive as used on the Jeffrey car. Both the bevel pinion and the differential housing which carries the driving gear or ring gear are carried by ball bearings. The action of the bevel gears

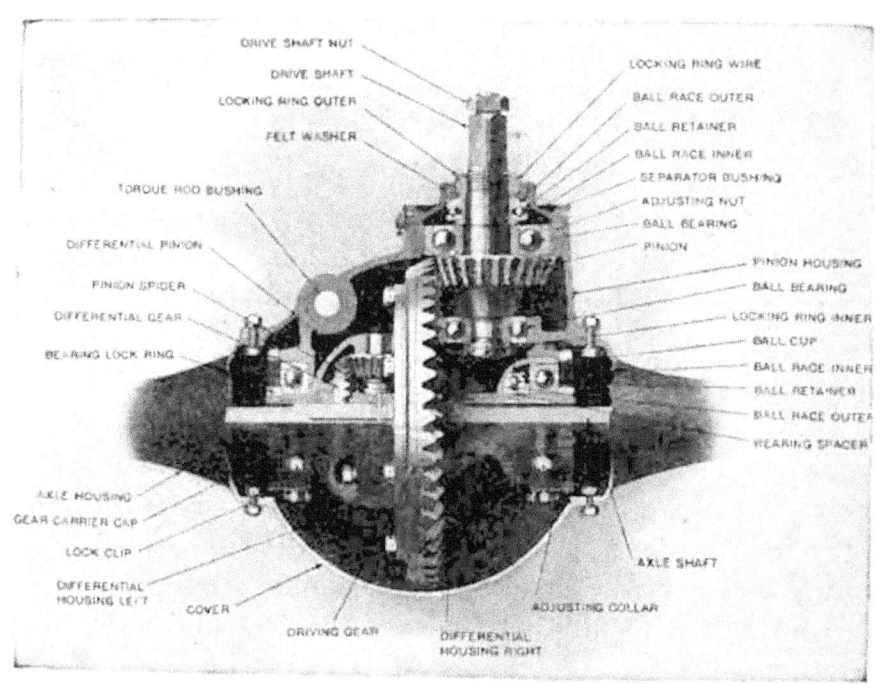

Fig. 87.—Jeffrey final drive.

produces a side thrust, caused by the inclination of the faces of the teeth, tending to separate the gears. This makes it necessary that the bearings of these gears be capable of resisting this thrust. Either ball bearings or tapered roller bearings are employed. If the straight rollers are used for bearings, special thrust bearings must be provided.

Figure 88 shows a spiral-bevel gear drive with the Timken tapered roller bearings, as used on the Cadillac car. The chief claims for the spiral-bevel drive are that the spiral teeth give a more continuous driving action between the teeth and overcome any possible inaccuracies in the teeth or any tendencies to wear irregularly; also that they overcome the thrust, to a more or less extent, by producing a counteracting pull.

Fig. 89 shows the worm drive to the rear axle. This has the worm placed above the gear. The worm drive in Fig. 59 shows one with the worm placed underneath. The worm drive is very quiet running, but requires careful lubrication because of the constant sliding action between the teeth of the worm and gear. One of the two gears should run in an oil bath. The worm drive is especially popular in

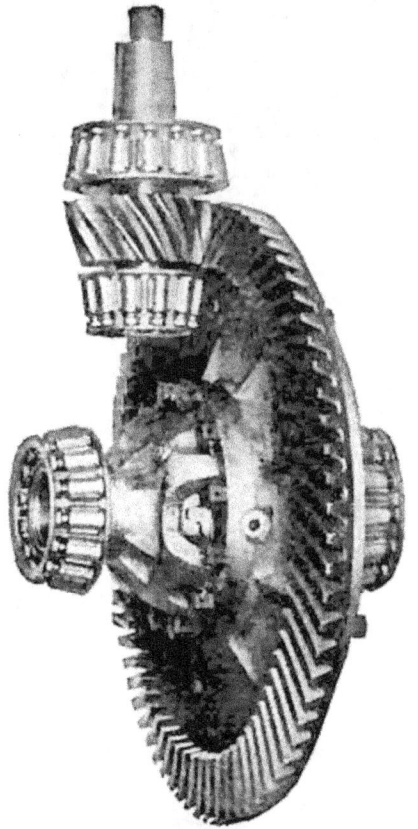

Fig. 88.—Cadillac spiral-bevel drive.

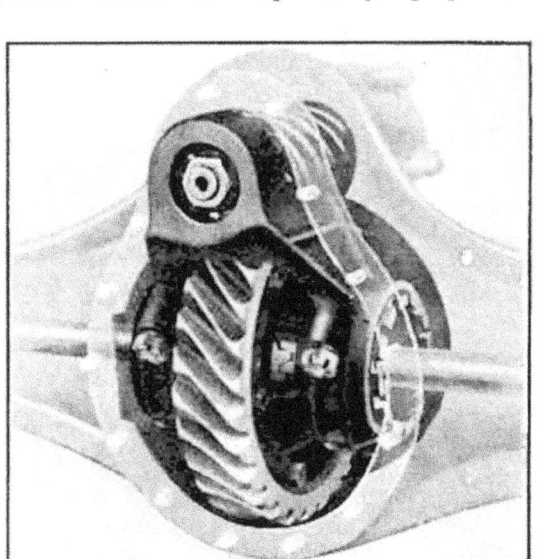

Fig. 89.—Worm drive used on Jeffrey "Chesterfield Six."

heavy truck service where there is a large reduction in speed. The worm is generally made of steel and the gear of bronze to keep down the friction.

49. Types of Live Rear Axles.—The dead rear axle was illustrated and explained in Chap. I. The live axle is used on practically all makes of pleasure cars, with only one or two exceptions. Live rear axles are classified according to their methods of construction as *simple, semi-floating, three-quarter floating,* and *full floating.*

Simple Live Axle.—The simple live axle used on the Ford Model T is shown in Fig. 90. This type of rear axle performs two functions in that it carries the entire weight of the rear of the car in addition to transmitting the power. The rear wheel is keyed to the axle as shown. The weight is carried by roller bearings directly on the live axles both at the wheel and differential ends.

Semi-floating Axle.—Figure 91 is of the semi-floating type and shows

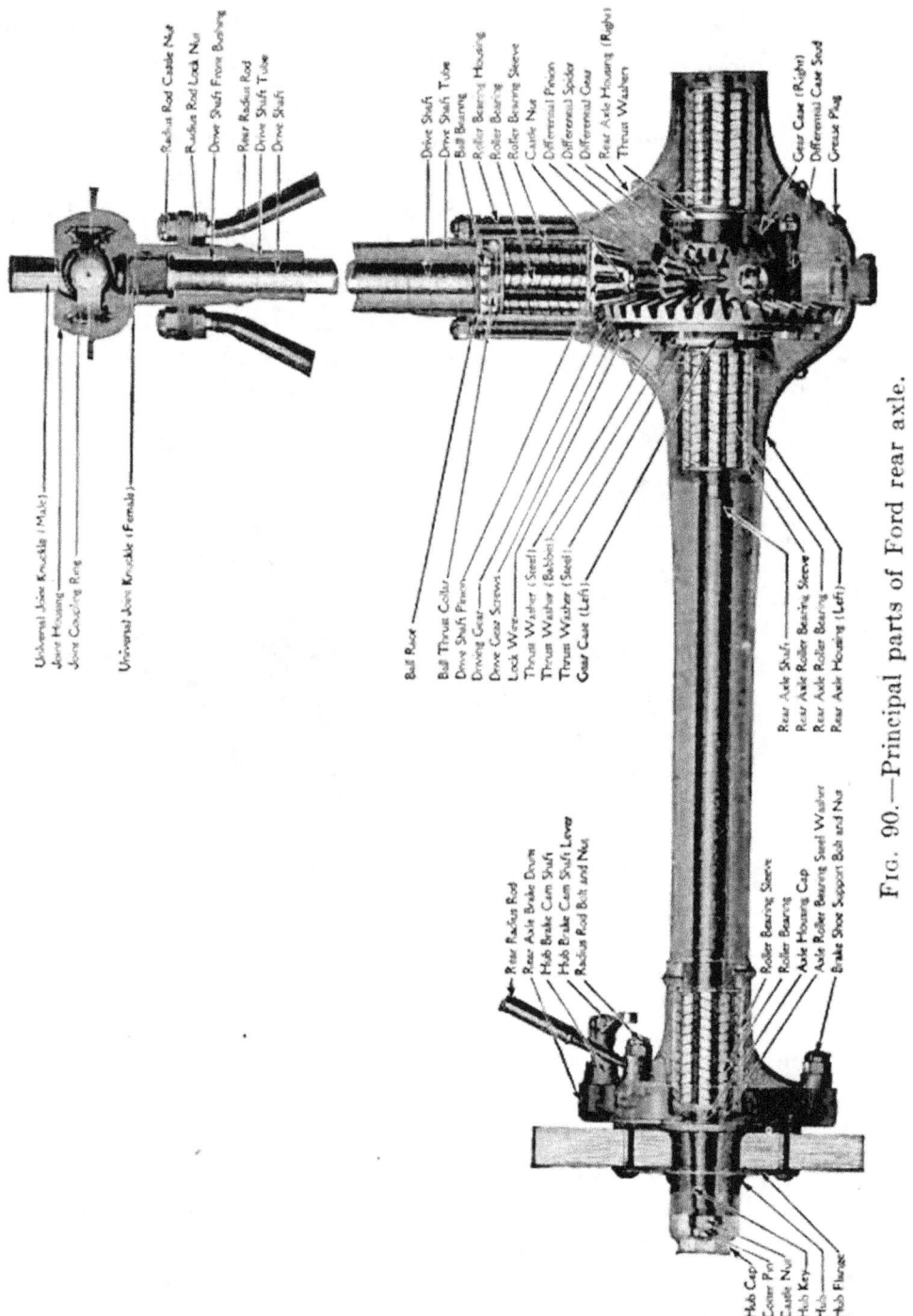

Fig. 90.—Principal parts of Ford rear axle.

the essential difference between a simple and semi-floating live axle. In the semi-floating axle the inner bearings are carried on an extension of the differential case, thus relieving this end of the live axle of considerable stress. The wheel as in the other case is keyed to the axle. The construction at the outer end of the semi-floating axle is the same as in the simple axle. In either of these types the weight of the car produces a bending stress in the axle.

Three-quarter Floating Axle.—Figure 92 shows the change in this type of construction from the semi-floating type. The weight is carried by the bearings on the housing and directly in line with the spokes, thus relieving the axle of all bearing stresses. The wheel is keyed onto the shaft.

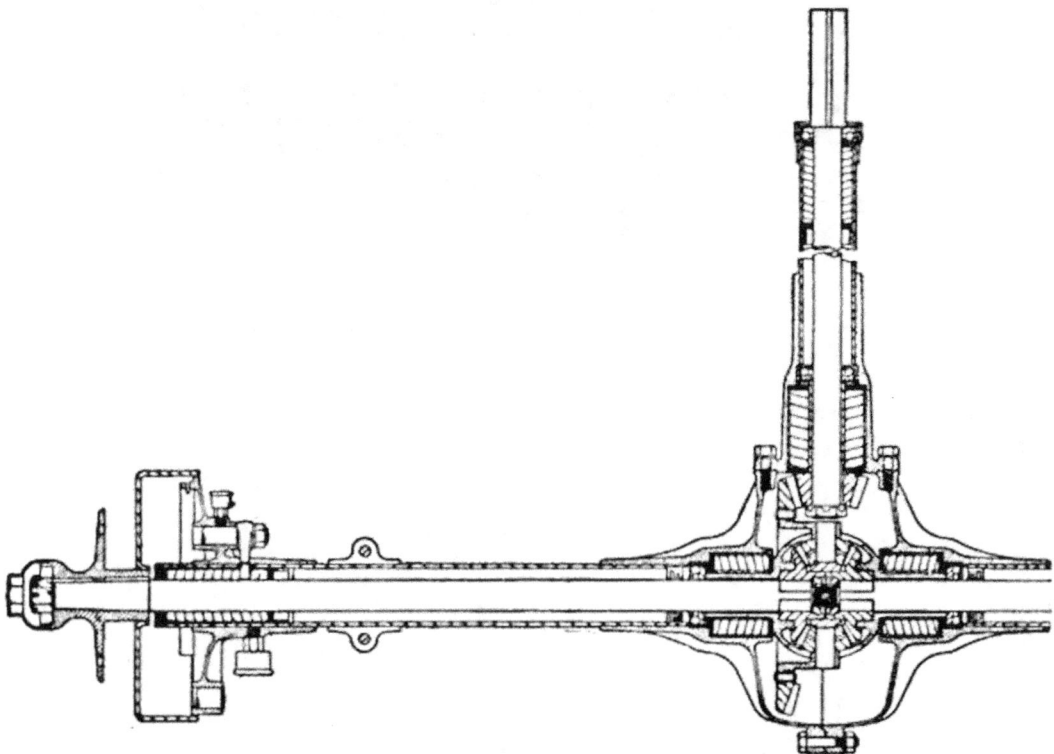

Fig. 91.—Semi-floating rear axle.

Although in the three-quarter type the live axle is relieved of all weight, nevertheless the bending strains due to a possible side movement of the wheel, or the distortion due to a bent housing are still thrown on the axle due to the fact that the wheel is keyed onto the axle. Also, in this type, if the live axle breaks, the wheel can come off and let the car drop. This is prevented only by the full-floating construction.

Full-floating Rear Axles.—Figure 93 shows the full-floating construction as used on the Buick car. The wheel is carried on a double ball or roller bearing on the axle housing, in such a way as to retain the wheel on the housing regardless of what may happen to the live axle. In this construction, the live shaft receives only the torsional strains of driving the car, all other loads being taken by the axle housing. The live shaft may be removed and replaced without disturbing either the wheel or the differential. The inner ends of the axle shafts are grooved and slide into corresponding grooves in the differential gears. The entire drive shaft on either side may be removed by merely removing a hub cap and sliding the shaft out. In the form shown in Fig. 93, the shaft is keyed into the

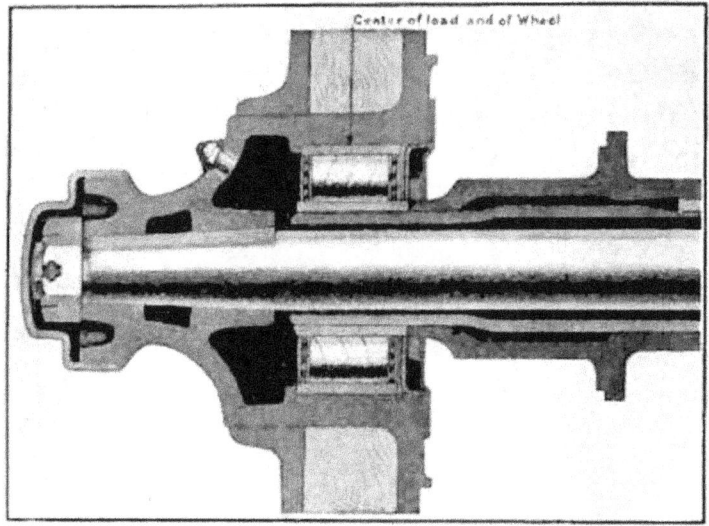

Fig. 92.—Three-quarter floating axle construction.

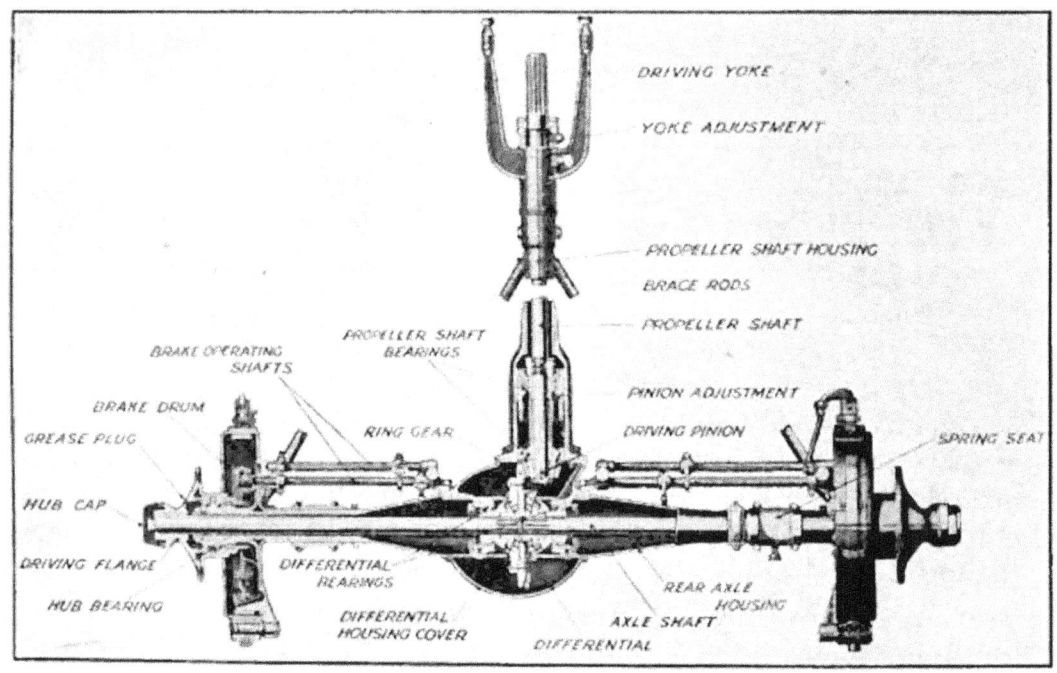

Fig. 93.—Buick full-floating rear axle.

hub cap. In another form, the outer end of the shaft has a toothed clutch which fits into corresponding recesses in the outer face of the hub. This permits a certain amount of play and relieves the shaft from any distortion if the axle housing becomes bent.

CHAPTER IV

FUELS AND CARBURETTING SYSTEMS

One of the most important operations in a gas engine is that of getting an explosive mixture inside of the engine cylinder at the proper time. This explosive mixture is formed by the thorough mixing of air and a gas formed by the evaporation of a volatile liquid fuel, usually gasoline.

50. Hydrocarbon Oils.—Most of the liquid fuels are known as "hydrocarbon" oils, because they are made from crude mineral oil containing as its principal parts, hydrogen and carbon. One of the hydrocarbon fuels, viz., alcohol, is not of mineral derivation, but is made by the distillation of vegetable matter.

The crude oil or petroleum from which the hydrocarbon fuels are made is found in natural deposits several hundred feet below the earth's surface. In some places it has to be pumped out, while in others it is forced out by natural gas pressure. Most of the crude oil found in the United States comes from Pennsylvania, Ohio, Illinois, Kansas, Texas, Oklahoma and California. These crude oils are of two general types, that coming from Texas, Oklahoma, and California having what is known as an "asphalt" base, and that from Pennsylvania and Ohio having a "paraffin" base. Crude oil having an "asphalt" base is a heavy dark liquid, which when boiled, leaves a black tarry residue. If the crude oil has a "paraffin" base, it is much lighter in weight and color and, when boiled, leaves a residue from which is made the white paraffin or wax with which everyone is familiar.

Formerly, gasoline made from crude oil with a paraffin base was supposed to be of a higher grade than the other, but with the modern processes of refining, the gasoline from the two kinds of crude oil gives equally good results.

51. Fractional Distillation of Petroleum.—The crude oil is heated in large retorts or "stills," provided with accurate temperature recording devices. When the temperature has reached about 100°F. a vapor begins to rise from the oil. This vapor is collected from the top of the retort and condensed in cooling coils, from which the liquid is collected in vessels. As the temperature in the retort rises, the vapor becomes heavier and, when condensed, gives the heavier and less volatile liquid fuels. The following table gives, approximately, the products of this method of distillation:

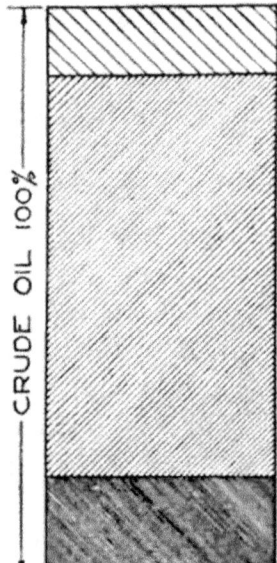

Fig. 94.—Approximate fractions in the distillation of crude oil.

Temperature in the retort	Kind of oil after condensing the vapor	Percentage
100°F to 125°F.	Highly volatile oils (gasoline, benzine and naphtha).	10 to 15 per cent.
125°F. to 350°F.	Kerosene and light lubricating oils.	65 to 75 per cent.
Over 350°F.	Heavy oils, paraffin wax and residue.	15 to 20 per cent.

It will be noticed that there is from three to five times as much kerosene and light lubricating oils produced under this method as there is gasoline. This accounts for the late scarcity of gasoline and the more volatile fuels, and the overproduction of kerosene and the less volatile fuels, which can not be used successfully in an automobile engine.

In order to utilize a part of these less volatile fuels, the Standard Oil Co. has developed the Burton process by which these less volatile fuels are redistilled under pressure. This process gives an additional amount of volatile fuel very much like the gasoline obtained from the first distillation. This process has increased the percentage of gasoline from the crude oil to such an extent that the market is now liberally supplied.

The Bureau of Mines has recently developed the new Rittman process for increasing the amount of gasoline produced from the crude oil. It is a continuous process, in contrast to the "batch" Burton process. The two processes are somewhat similar in character and have as their end an increase in the production of gasoline from the crude oil.

52. Principles of Vaporization.—Before an explosive mixture can be formed, the liquid fuel must first be turned into a gas and then mixed with the proper amount of air to burn it. As we know, it requires heat to change water into steam or vapor. If the water is out in the open, it will evaporate rapidly, or boil, at a temperature of 212°. Likewise, in order to change a liquid fuel into a gas or vapor, it is necessary that heat be added to it, but the temperature at which this heat is added is different for different fuels. For instance, gasoline will evaporate under the usual atmospheric pressure and temperature and will, in some cases, evaporate at a temperature close to 0° F. This can be tested by exposing a pan of gasoline to the air. In a short time the liquid will have evaporated. That heat has been absorbed can be verified by feeling of the dish before it is filled and again after evaporation has been taking place.

Kerosene and alcohol, on the other hand, will not evaporate until heat is added from an external source at a higher temperature, the same as is done when steam is made from water. This explains the difficulty of evaporating these fuels for use in a gas engine.

From the above considerations, the general principles of vaporization are formulated:

1. The heavier a liquid and the higher its boiling point, the harder it will vaporize; for example, kerosene as compared with gasoline.

2. A liquid fuel will vaporize easier and faster under a suction, or reduction of pressure than under pressure; for example, gasoline is more difficult to vaporize at low than at high altitudes.

3. The closer the temperature of a liquid fuel is to its boiling point, the easier and faster it will vaporize; for example, gasoline will vaporize more readily in summer than in winter.

The Baumé Test.—Gasoline is usually spoken of as *high* or *low* test. By reference to the principles of vaporization, we see that the heavier a liquid, the harder it is to evaporate. This principle explains the reason for the use of the Baumé test. A hydrometer, such as shown in Fig. 95 is graduated in degrees, the numbers reading from the bottom up. These degrees have nothing to do with thermometer degrees, but are named after Baumé, who originated the idea. When the hydrometer is placed in a quantity of gasoline, it will sink to a depth corresponding to the density of the liquid. It will sink deeper in a light gasoline than in a heavier one. The deeper the hydrometer sinks, the higher the scale reading will be. This scale, reading from 45 to 95° Baumé, indicates in an indirect way the ease and rapidity with which the gasoline will evaporate. It is not a direct and absolute test unless the nature and the boiling points of the crude oil from which the gasoline has been distilled are known. For most purposes, however, it merely serves as a guide as to the way the gasoline will act in service.

Gasoline.—The commercial gasoline of today has a Baumé test of from 50 to 65°, the better or high test being in the neighborhood of 65° and the poorer, or low test, in the neighborhood of 50°. For summer use, the low test or heavier gasoline can be used very well because it will evaporate with comparative ease at the usual summer temperatures, but in the winter the high test or light gasoline is to be preferred because it will evaporate more easily at the low temperatures. More work can be obtained from a gallon of the heavier or low test gasoline, providing it is completely vaporized, but it is very difficult to vaporize at low temperatures and consequently makes starting very hard in cold weather.

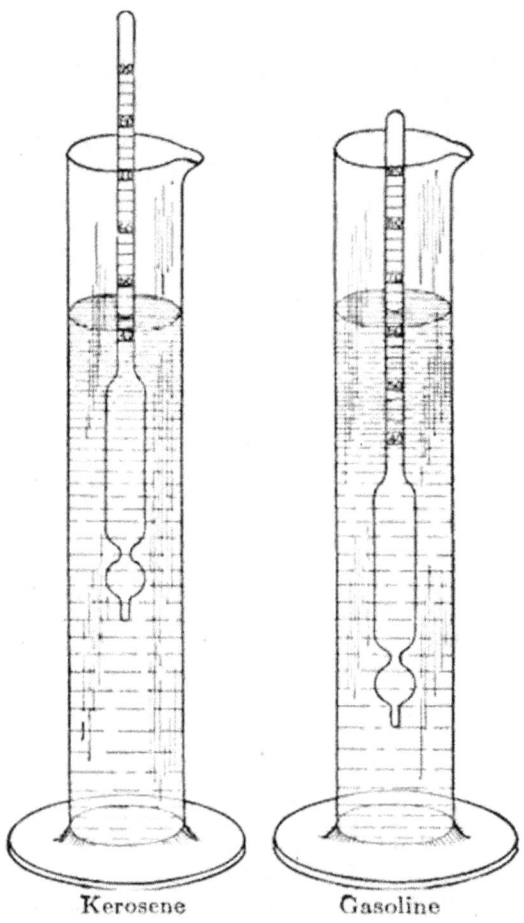

Fig. 95.—Baumé hydrometer in kerosene and gasoline.

Occasionally, a low grade, impure gasoline is sold which lacks sufficient refinement and purification, the sulphur and other impurities not having been eliminated. The use of this may result in carbon deposits in the cylinders. A gasoline that readily carbonizes should be avoided and a higher grade used.

Kerosene and Alcohol.—To use either of these fuels requires the heating of the fuel or the air, or both, in order to secure vaporization. At present, the price of alcohol is too high to warrant giving any serious consideration to its use. Several more or less successful devices have been tried for using kerosene, but the varying speeds and loads of the automobile engine make the problem of controlling the heat very difficult. The reductions in the price of gasoline in the past 2 or 3 years and the very promising prospects for a greater increase in the supply and corresponding reduction in the price, make it unlikely that any great development in the use of kerosene will take place. Consequently, the discussion to follow will deal only with gasoline and its vaporization.

53. Heating Value of Fuels.—The heating value, or the amount of heat energy contained in a liquid fuel, is given in British thermal units per *pound*; a British thermal unit, or a B.t.u., being the quantity of heat energy required to raise the temperature of 1 lb. of water 1° on the Fahrenheit scale. The following table gives the heating values of the common fuels:

Gasoline 18,000 to 19,500 B.t.u. per pound.
Kerosene about 20,000 B.t.u. per pound.
Alcohol { grain about 10,000 B.t.u. per pound.
{ wood about 7,500 B.t.u. per pound.

Inasmuch as the heavier fuel contains more pounds per gallon, and as gasoline and kerosene are sold by the gallon, a gallon of heavy or low test gasoline or of kerosene contains more energy and gives more power than a gallon of light, or high test gasoline.

54. Gasoline Gas and Air Mixtures.—It is necessary when the gasoline is vaporized that it be mixed with the proper amount of air to form an explosive mixture. If too little air is furnished, there will not be enough oxygen to burn the carbon and hydrogen in the fuel and the fuel will be wasted, as will be indicated by black smoke coming from the exhaust. If too much air is furnished, the mixture is weak in fuel, giving a very slow combustion. This results in lost power. A weak mixture, or an excess of air, is indicated by back-firing through the carburetor.

A definite mixture of gasoline gas and air is necessary for the efficient operation of a gasoline engine. The function of the carburetor is to take the gasoline, vaporize it, and furnish the proper mixture of gas and air to the cylinders under all conditions of temperature, speed, load, power and varying atmospheric conditions.

55. Principles of Carburetor Construction.—Most of the modern types of carburetors are of the spray or nozzle type, in which a jet of gasoline is sprayed into a current of air to form an explosive mixture. Figure 96 illustrates an elementary spray carburetor. The gasoline supply tank is placed below the carburetor and the gasoline is pumped up through the supply pipe. The overflow pipe maintains the level of the liquid at a constant height. The standpipe T is connected with the supply chamber C by means of the connection N and the flow is regulated by the needle valve S. The gasoline level in the standpipe T is always the same. The flange B is fastened onto the intake passage of the engine. The suction of the piston draws air through the opening A upward past the standpipe, and at the same time draws a spray of gasoline from T. The butterfly valve D is for the purpose of regulating the suction upon the standpipe T when starting the engine; when running, the valve D should be wide open. The mixture is changed by regulating the needle valve S. This type of carburetor can be used only on constant speed engines, the reason for which we will see later. Figure 97 shows another elementary type of carburetor which illustrates the application of two modern ideas. In this case, the gasoline supply is maintained at a constant level by means of a hollow metal or a cork float operating a ball valve. The

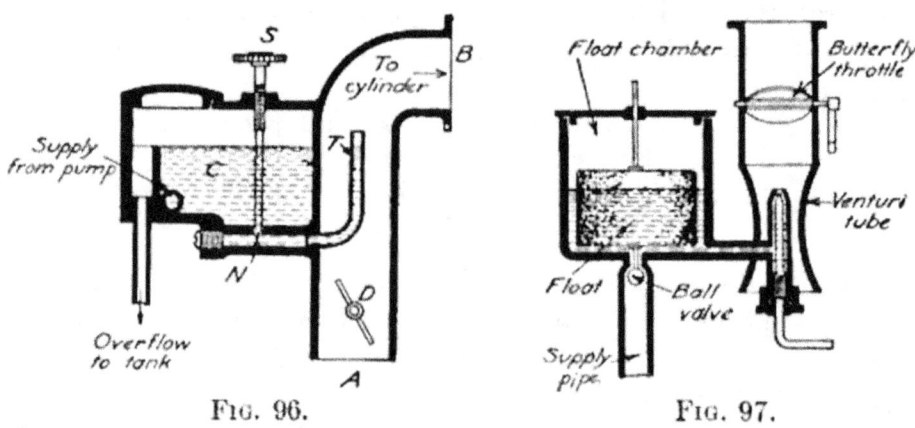

Fig. 96.   Fig. 97.

arrangement requires the gasoline supply tank to be placed above the carburetor or that some other means be provided for supplying gasoline under pressure. It will also be noticed that the passage surrounding the standpipe or spray nozzle is contracted, giving the inside surface a convex shape. This is the application of the well known Venturi tube principle. By contracting the section near the opening of the nozzle the velocity of the incoming air and consequently the suction at that point are increased, making it much easier for the gasoline to be taken up and greatly facilitating the starting of an engine when the suction is low.

This type of carburetor could be used on constant speed engines only. If a carburetor such as shown in Figs. 96 or 97 was put on a variable speed engine and the proper adjustment made by means of the needle valve so that the mixture proportions were correct at low speed, and the engine should then be speeded up, we would discover black smoke coming from the exhaust, indicating an excess of gasoline over the air supplied. This is due to the fact that under the increased suction due to the higher speeds of the piston, the air drawn in past the standpipe expands and increases in volume and velocity faster than it increases in weight; while the gasoline drawn from the nozzle, being a liquid, increases in weight just as its velocity and volume are increased. This means that under an increased suction too much gasoline is supplied for the amount of air drawn in.

FUELS AND CARBURETTING SYSTEMS

In order to keep the mixture of the proper proportions at all speeds of the engine, it is necessary to have an auxiliary air entrance, such as indicated at X in Fig. 98, to admit an additional amount of air at the higher engine speeds. This entrance is usually in the form of a valve controlled by a spring, the tension on which can be changed to control the air admission. For low speed adjustments the gasoline needle valve is to be used, and for high speed adjustments the auxiliary air valve is to be adjusted. That is, when the engine is running comparatively slowly, the air is taken in through the ordinary air opening A shown below the valve in Fig. 98.

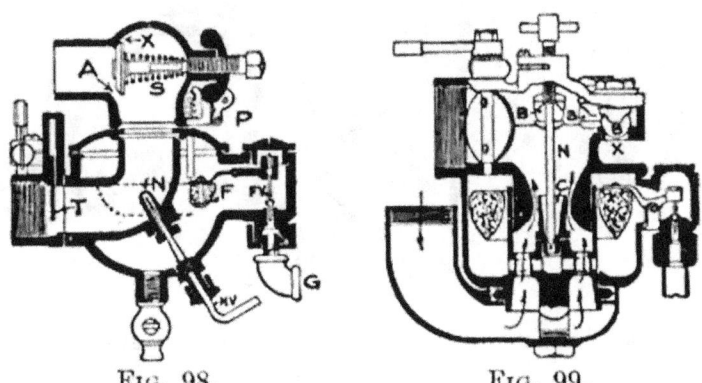

FIG. 98. FIG. 99.
FIGS. 98 AND 99.—Sections of typical variable speed carburetors.

The mixture is then proportioned by means of the needle valve NV. When the engine speeds up, and the suction is increased, the auxiliary air valve S in Fig. 98 comes into action and opens. If it is found that the mixture at high speeds is too rich, that is if there is too much fuel for the air furnished, it indicates that the tension on the valve spring is too great, which prevents sufficient air from entering. By reducing the tension, the valve opens wider, letting in sufficient air to keep the mixture uniform. If the mixture is too weak at high speeds, the spring tension is too weak. It should be tightened so as to permit less air to enter and to increase the suction on the gasoline.

The following general description applies to Fig. 98.

G = gasoline feed from tank.
FV = float valve controlling flow of gasoline to carburetor.
F = float, the height of which is regulated by the level of gasoline in the float chamber. The float controls the float valve FV.
NV = gasoline needle valve for regulating the amount of gasoline furnished to the air in the mixing chamber.
N = gasoline nozzle.
X = auxiliary air valve, to admit additional air at the high speeds.
S = spring for X.
A = primary air opening, which supplies all air at low speeds.
T = throttle valve for regulating supply of mixture from carburetor to cylinder.
P = primer for depressing float and flooding carburetor to insure rich mixture when starting.

Figure 99 shows another carburetor, in which the auxiliary air is admitted through ports X controlled by steel balls B.

Some of the modern types of carburetors are water-jacketed, taking the hot water from the cooling system, in order to heat the carburetor and assist the vaporization. Another method of assisting the vaporization, and one almost necessary when the low grade gasoline of today is considered, is that of heating the air which goes into the carburetor. This is usually done by taking it through a jacket surrounding the exhaust pipe. Figure 100 shows such a device.

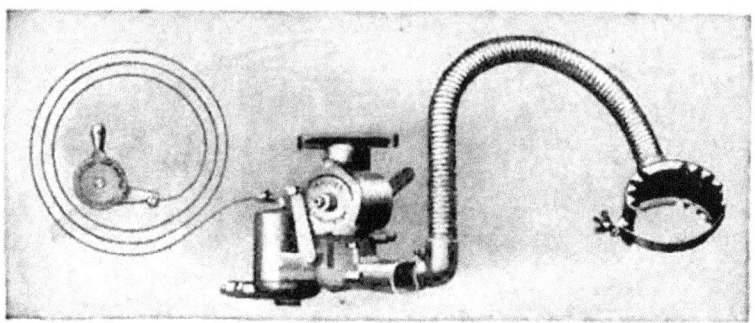

Fig. 100.—Hot-air connection used with Master carburetor.

Another scheme used in several of the carburetors built for high powered, high speed machines is the double-jet, which makes it easier for the engine to draw the desired amounts of gasoline and air when it becomes necessary for the engine to carry heavy loads at high speed. Several of these are illustrated in the following articles, which describe some of the leading carburetors now in use.

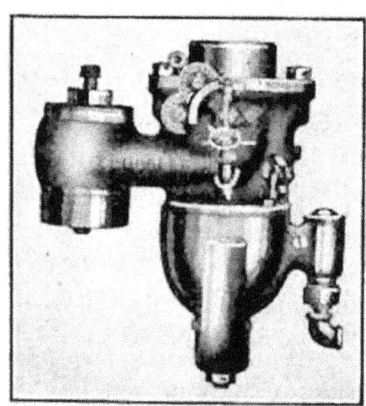

Fig. 101.—Schebler Model L carburetor.

56. Schebler Model L Carburetor.—The Model L carburetor, Figs. 101 and 102, is of the lift-needle type and is so designed that the amount of fuel entering the motor is controlled by means of a raised needle working automatically with the throttle. The flow of gasoline can be adjusted for closed, intermediate, or open throttle positions, each adjustment being independent and not affecting either of the others. This carburetor has an automatic air valve, shown at the left in Fig. 102. At high speeds or heavy loads, the suction raises this valve and admits an extra supply of air. The opening of the throttle for high speed or a heavy pull raises the needle and increases the supply of gasoline to correspond with the increased air supply.

The Model L can be furnished with a bend for connecting or taking warm air from around the exhaust manifold into the initial air opening at the base of the carburetor, by means of a hot air drum and tubing.

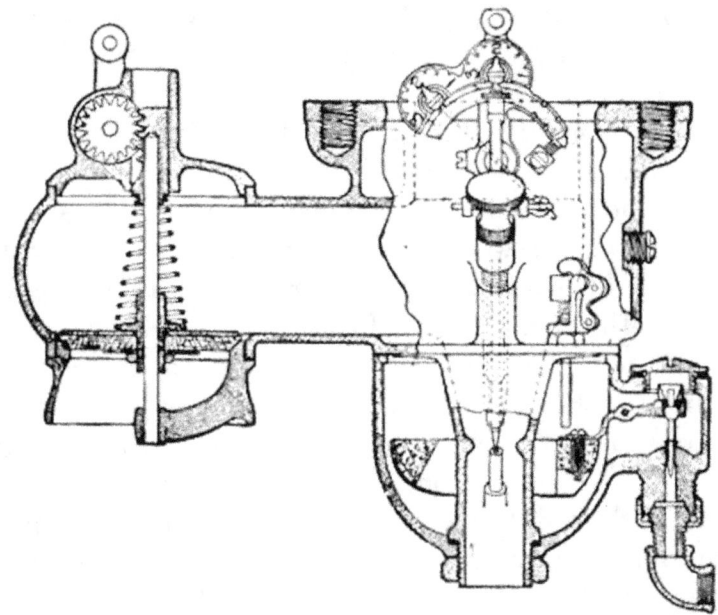

Fig. 102.—Section of Shebler Model L carburetor.

This carburetor is also manufactured with a dash-control to the air valve spring, this being operated by a lever which is controlled by a switch on the dashboard or steering post of the car. This control is shown on Fig. 102.

Rules for Adjusting Schebler Model L.—The carburetor should be connected to the intake manifold so that it is located below the bottom of the gasoline tank a sufficient distance to be filled by gravity under all running conditions. Where pressure feed is used, it is unnecessary to locate the carburetor below the gasoline tank; also, when pressure is used, it is never advisable to carry over 2 lb.

Before adjusting the carburetor, make sure that the ignition is properly timed; that there is a good hot spark at each plug; that the valves are properly timed and seated; that all connections between the intake valves and the carburetor are tight; and that there are no air leaks of any kind in these connections. The carburetor should be adjusted to the motor under normal running temperature, and not to a cold motor.

In adjusting the carburetor, first make the adjustments on the auxiliary air valve so that the air valve seats lightly but firmly. The lever on the dash control should be set in the center of the dashboard adjuster, and with this setting of the lever, the tension on the air valve should be light, yet firm. Close the needle valve by turning the adjustment screw to the right until it stops. Do not use any pressure on this adjustment screw after it meets with resistance. Then turn it to the left about four or five turns and prime or flush the carburetor by pulling up the priming lever and holding it up for about 5 seconds. Next, open the throttle about one-third and start the motor; then close the throttle slightly, retard the spark, and adjust the throttle lever screw and the needle valve adjusting screw, so that the motor runs at the desired speed and hits on all cylinders.

After getting a good adjustment with the motor running idle do not touch the needle valve adjustment again, but make your intermediate and high speed adjustments on the dials. Adjust the pointer on the first dial about half way between figure 1 and figure 3. Advance the spark and open the throttle so that the roller on the track running below the dials is in line with the first dial. If the motor back-fires with the throttle in this position and the spark advanced, turn the indicator a little more toward figure 3; if the mixture is too rich, turn the indicator back, or toward figure 1, until you are satisfied that the motor is running properly with the throttle in this position, or at intermediate speed. Now open the throttle wide and make the adjustment on the second dial for high speed in the same manner as you have made the adjustments for intermediate speed on the first dial.

57. Schebler Model R.—The Model R Schebler carburetor, Fig. 103, is designed for use on both four- and six-cylinder motors. It is a single-jet raised-needle type of carburetor, automatic in action. The air valve controls the lift of the needle so as to automatically proportion the amount of gasoline and air at all speeds.

The Model R carburetor is designed with an adjustment for low speed. As the speed of the motor increases, the air valve opens, raising the gasoline needle and thus automatically increasing the amount of fuel. This carburetor has but two adjustments, the low speed needle adjustment. which is made by turning the air valve cap, and an adjustment on the air valve spring for changing its tension.

The Model R carburetor has an eccentric which acts on the needle valve, intended to be operated either from the steering column or from the dash, and insures easy starting, as, by raising the needle from the seat, an extremely rich mixture is furnished for starting and for heating up the motor in cold weather. A choke valve in the air bend is also furnished. The dashboard control or steering column control must be used with this carburetor; it cannot be operated satisfactorily without them.

Rules for Adjusting Model R Carburetor.—When the carburetor is installed, see that lever B is attached to the steering column control or dash control so that when boss D of lever B is against stop C the lever on the steering column control or dash control will register "Lean" or "Air." This is the proper running position for lever B.

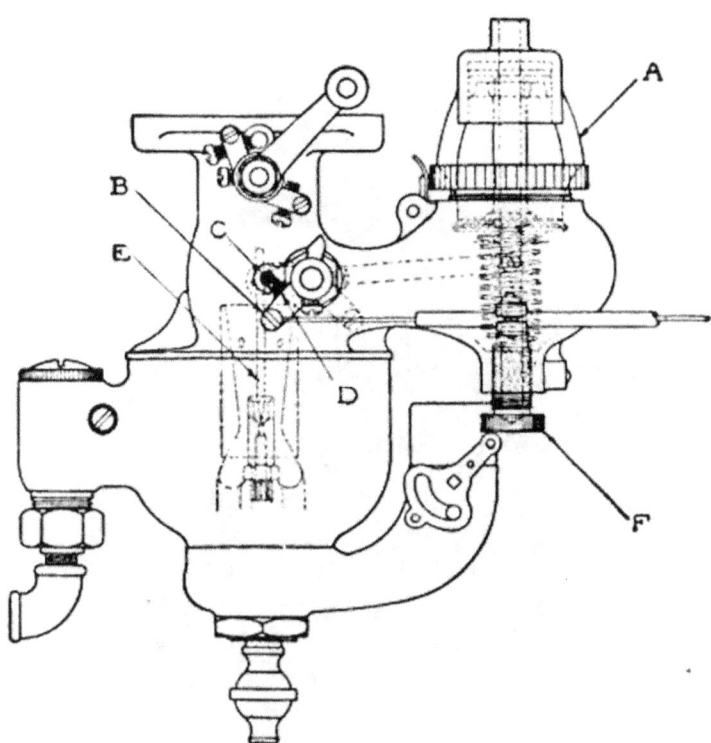

Fig. 103.—Schebler Model R carburetor.

To adjust the carburetor turn the air valve cap A clockwise or to the right until it stops, then turn it to the left or anti-clockwise one complete turn.

To start the engine, open the throttle about one-eighth or one-quarter way. When the engine is started, let it run till it is warmed, then turn the air valve cap A to the left or anti-clockwise until the engine hits perfectly. Advance the spark three-quarters of the way on the quadrant; then if the engine back-fires on quick acceleration, turn the adjusting screw F up (which increases the tension on the air valve spring) until acceleration is satisfactory.

Turning the air valve cap A to the right or clockwise lifts the needle E out of the nozzle and enriches the mixture; turning to the left or anti-clockwise lowers the needle into the nozzle and makes the mixture lean.

When the motor is cold or the car has been standing, move the steering column or dash control lever toward "Gas" or "Rich." This operates the lever B and lifts the needle E out of the gasoline nozzle and makes a rich mixture for starting. As the motor warms up, move the control lever gradually back toward "Air" or "Lean" to obtain best running conditions, until the motor has reached normal temperature. When this temperature is reached, the control lever should be at "Air" or "Lean." For best economy, the slow speed adjustment should be made as lean as possible.

58. The Holley Model H Carburetor.—This carburetor is shown in Fig. 104. Before the fuel enters the float chamber, it passes a strainer

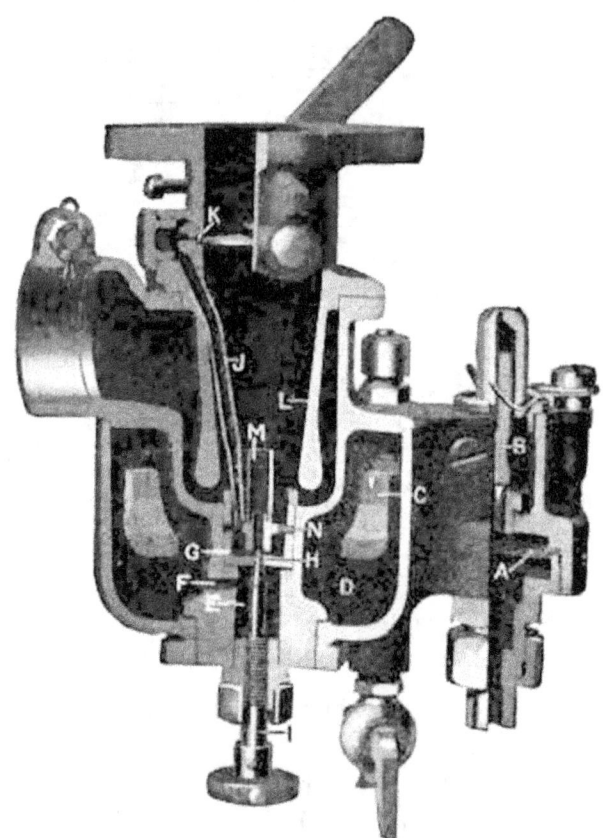

Fig. 104.—Holley Model H carburetor.

disc A which removes all foreign matter that might interfere with the seating of the float valve B under the action of the cork float, and its lever C.

Fuel passes from the float chamber D into the nozzle well E through a passage F drilled through the wall separating them. From the nozzle well, the fuel enters the cup G through the hole H, and rises past the needle valve, I, to a level which partially submerges the lower end of a small tube, J, having its outlet K at the edge of the throttle disc.

Cranking the engine, with the throttle kept nearly closed, causes a very energetic flow of air through the tube J and its calibrated throttling plug K. But with the engine at rest the lower end of this tube is partially submerged in fuel. Therefore, the act of cranking automatically primes the motor. With the motor turning over under its own power, flow through the tube J takes place at very high velocity, thus causing the fuel entering the tube with the air to be thoroughly atomized upon its exit from the small opening at the throttle edge. This tube is called the "low speed tube" because, for starting and idle running, all of the fuel and most of the air in the working mixture are taken through it.

As the throttle opening is increased beyond that needed for idling of the motor, a considerable volume of air is drawn down around the outside of the strangling tube L and then upward through this tube. In its passage into the strangling tube, the air is made to assume an annular, converging stream form, so that the point in its flow at which it attains its highest velocity is in the immediate neighborhood of the upper end of the "standpipe" M. The velocity of air flow being highest at the upper or outlet end of the standpipe, the pressure in the air stream is lowest at the same point. For this reason, there is a pressure difference between the top and bottom openings of the pipe M, thus causing air to flow through it from bottom to top, the air passing downward through the openings N in the bridge supporting the standpipe and then up through the standpipe.

With a very small throttle opening, the action through the standpipe keeps the nozzle cup thoroughly cleaned out, the fuel being carried directly from the needle opening into the entrance of the standpipe. To secure the best vaporization of the fuel, the passage through the standpipe is given an aspirator form, which further increases the velocity of flow through it, and insures the greatest possible mixing of the fuel with the air. A further point is that the vaporized discharge of the standpipe enters the main air stream at the point at which the latter attains its highest velocity and lowest pressure.

There is but one adjustment, that of the needle valve I. The effect of a change in its setting is manifest over the whole range of the motor.

59. Holley Model G.—This carburetor, Fig. 105, is a special design for Ford cars.

The operation of this carburetor is the same as the regular Model H already illustrated and described. The chief differences are the structural ones giving a horizontal instead of a vertical outlet, a needle valve controlled from above instead of from below, and a simplification of design to secure compactness.

Fuel enters the carburetor by way of a float mechanism in which a hinged ring float, in rising with the fuel, raises the float valve into contact with its seat. This seat is removable and the float valve is provided with a tip of hard material.

From the float chamber the gasoline passes through the ports E to the nozzle orifice, in which is located the pointed end of the needle F. The ports E are well above the bottom of the float chamber, so that, even should water or other foreign matter enter the float chamber, it would have to be present in very considerable quantity before it could interfere with the operation of the carburetor. A drain valve D is provided for the purpose of drawing off whatever sediment or water may accumulate in the float chamber.

Fig. 105.—Holley Model G carburetor.

The float level is so set that the gasoline rises past the needle valve F and sufficiently fills the cup G to submerge the lower end of the small tube H. Drilled passages in the casting communicate the upper end of this tube with an outlet at the edge of the throttle disc. The tube and passage give the starting and idling actions, as described in connection with the Holley Model H.

The strangling tube I gives the entering air stream an annular converging form, in which the lowest pressure and highest velocity occur immediately above the cup G; thus it is seen that the fuel issuing past the needle valve F is immediately picked up by the main air stream, at the point of the latter's highest velocity.

The lever L operates the throttle in the mixture outlet. A larger disc with its lever S forms a spring-returned choke valve in the air intake, for starting in extremely cold weather.

60. Stewart Model 25.—This carburetor, which is manufactured by the Detroit Lubricator Company, involves an interesting principle of operation.

Figure 106 gives a cross section of this carburetor and shows the position of the air valve with engine running and air and gasoline being admitted.

With the engine at rest and no air passing through the carburetor, the air valve A rests on the seat B, closing the main air passage. The gasoline rises to a height of about $1\frac{1}{2}$ in. below the top of the central aspirating

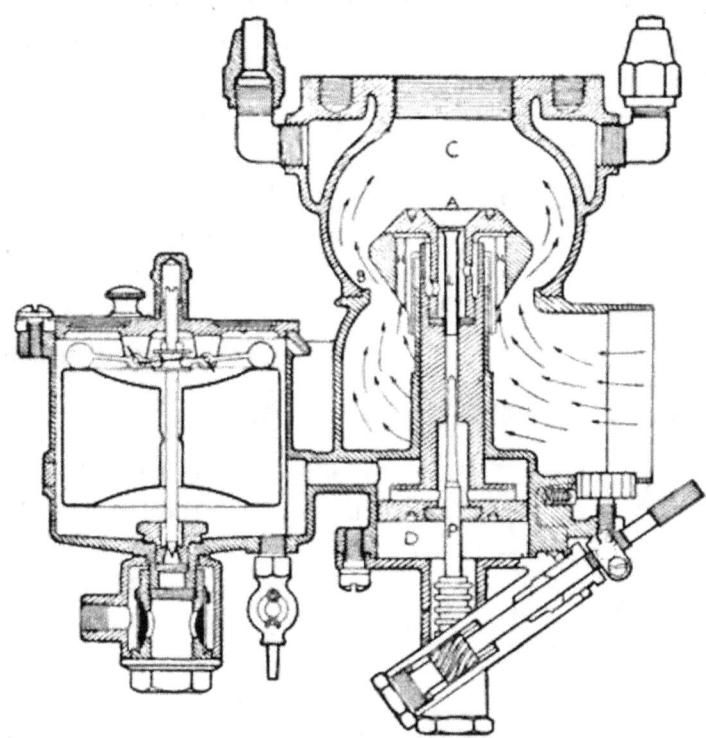

Fig. 106.—Stewart Model 25 carburetor.

tube L. As soon as the engine starts to rotate, a partial vacuum is formed above the air valve, causing it to lift from its seat and admit air, at the same time gasoline being drawn up through the aspirating tube L. The lower end of the air valve extends down into the gasoline and around the metering pin P. Due to the decreasing diameter of this pin, the higher the air valve is lifted the larger will be the opening into the tube L, and the more gasoline will be drawn up. The upper end of the air valve measures the air, the lower end measures the gasoline; therefore, as the suction varies, the air valve moves up or down and the volume of air and the amount of gasoline admitted to the mixing chamber increase or decrease in the same ratio. Most of the air passing through the carburetor goes through the air passages as indicated by the black arrows. A small amount is drawn through the drilled holes HH and past the end of the tube L. The flared end of this tube deflects the air through a small annulus, thereby increasing the velocity of air at this point so as to aid in atomizing the fuel.

The air valve is restrained from any tendency to flutter, caused by the intermittent suction of the cylinders, by the dash pot D. Due to the greater inertia of the gasoline and because it flows comparatively slowly through the small opening and into the dash pot, the air valve can rise or fall only as liquid is expelled or admitted and thus the air valve is held steady. The Stewart carburetors have but one adjustment, which raises or lowers the metering pin, thereby decreasing or increasing the amount of gasoline admitted to the mixing chamber. The correct position of the metering pin is determined with the motor running at idling speed. This adjustment may be manipulated at the dash to compensate for extreme changes in atmospheric temperatures and for use in starting in cold weather.

61. Kingston Model L.—Figure 107 shows the construction of this carburetor. Gasoline enters the carburetor from the tank at the connection A and is maintained at a constant level, through the agency of the float.

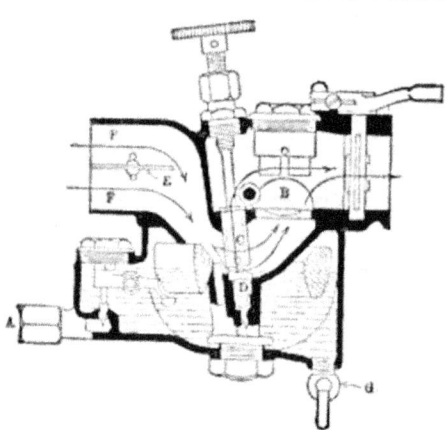

FIG. 107.—Kingston Model L carburetor.

A pool of gasoline forms in the base of the U-shaped mixing tube and will always be present when the motor is not running. This aids in positive starting. When the motor starts, this pool is quickly lowered to the point of adjustment of the needle valve and continues to feed from this point till the motor is stopped.

When the motor is running slowly, the air valve B rests lightly on its seat, allowing no air to pass through; consequently all air must pass through the low speed mixing tube C. Due to the lower end of this tube being close to the spray nozzle and all the low speed air having to pass this point, the atomized gasoline drawn from nozzle D becomes thoroughly mixed with air in its upward course and is carried in this state to the motor.

When the throttle is opened slowly, the following action takes place. The motor now requires a greater volume of mixture. The air valve B slowly leaves its seat, permitting an extra air supply to enter and compensate for the increased flow of gasoline produced by the greater suction of the motor. In this carburetor the extra amount of gasoline for the starting and warming up period can be obtained by opening the needle valve adjustment at the dash or by the use of the choke throttle E placed in the air passage.

When starting with a cold motor, this choke throttle can be closed by pulling the wire forward. This cuts off nearly all the air supply and produces a very strong suction at the spray nozzle, which causes the gasoline to jet up and be carried with the incoming rush of air to the cylinders.

A drain cock G is placed at the lowest point in the bowl and should be opened from time to time to free the bowl of all water and foreign matter.

Rules for Adjusting Kingston Model L.—Retard the spark fully. Open the throttle about five or six notches of the quadrant on the steering post. Loosen the needle valve binder nut on the carburetor until the needle valve turns easily. Turn the needle valve (with dash adjustment) until it seats lightly. Do not force it. Adjust it away from its seat one complete turn. This will be slightly more than necessary but will assist in easy starting.

Start the motor and open or close the throttle until the motor runs at fair speed, not too fast, and allow it to run long enough to warm up to service conditions. Now make the final adjustment. This carburetor has but one adjustment—the needle valve. Close the throttle until the motor runs at the desired idling speed. This can be controlled by adjusting the stop screw in the throttle lever.

Adjust the needle valve toward its seat slowly until the motor begins to lose speed, thus indicating a weak or lean mixture. Now adjust the needle valve away from its seat very slowly until the motor attains its best and most positive speed. This should complete the adjustment. Close the throttle until the motor runs slowly, then open it rapidly. The motor should respond strongly. Should the acceleration seem slightly weak or sluggish, a slight adjustment of the needle valve may be advisable to correct this condition. With the adjustment completed, tighten the binder nut until the needle valve turns hard.

62. Marvel Carburetor.—The Marvel, shown in Fig. 108, is of the double nozzle type, the second nozzle coming into action at high speeds. At low speeds all the air is drawn through the venturi tube, where it takes up gasoline from the primary nozzle. At high speeds after the air has passed the choke damper, it divides, part of it going through the venturi tube around the low speed spray nozzle, and the remainder passing to one side and opening the auxiliary valve against the pressure of its spring. Near the top of the auxiliary air valve is the secondary or high speed spray nozzle.

The rush of air through the venturi tube picks up and vaporizes the gasoline from the low speed nozzle and carries it in suspension past the throttle and to the cylinders. When the suction at the auxiliary air valve has increased sufficiently to open this valve and create a high velocity at this point, gasoline is also picked up from the high speed nozzle and carried to the cylinders in like manner.

The choke damper in the air inlet is used only for starting the motor, by partially shutting off the air supply and forcing the motor to suck in a rich mixture.

To the throttle is connected a hot air damper, which when open allows the exhaust gas from the motor to flow through a cored passage around the throttle, where it heats the mixture of gasoline and air. A tube connects this passage with another which surrounds the venturi tube and spray nozzle, and provides heat for the incoming fuel and air.

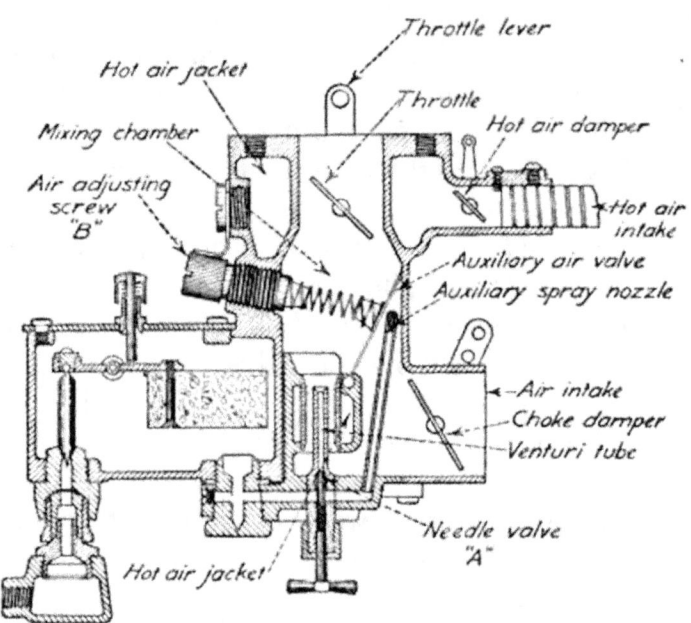

FIG. 108.—The Marvel carburetor.

Rules for Adjusting the Marvel Carburetor.—The following rules for adjustment are given by the manufacturers:

Start by turning the needle valve A to the right until it is completely closed, and the air adjustment B to the left until it stops. Now give the air adjusting screw B three complete turns to the right, and open the needle valve A two complete turns to the left. Start the motor as usual, using the strangler button to get a rich mixture at first. Close the throttle until the motor runs slowly and verify the needle valve adjustment A by turning it to the right a little at a time ($\frac{1}{2}$ to $\frac{3}{4}$ of a turn should be sufficient) until the motor runs smoothly and evenly. At this point the motor should be allowed to run until thoroughly warmed up.

After the motor has warmed up, turn the air valve adjusting screw B to the left, a little at a time, until the motor begins to slow down. This indicates that the air valve spring is too loose. Turn it back to the right just enough to make the motor run well.

To test the adjustment, advance the spark and open the throttle quickly. The motor should "take hold" instantly and speed up at once.

If it misses or pops back in the carburetor, open needle valve A slightly by turning to the left. Do not move the air adjusting screw B any more unless it appears absolutely necessary.

The best possible adjustment has been secured when the air adjustment B is turned as far as possible to the left and the needle valve A is turned as far as possible to the right, providing the motor runs smoothly and picks up quickly when the throttle is opened.

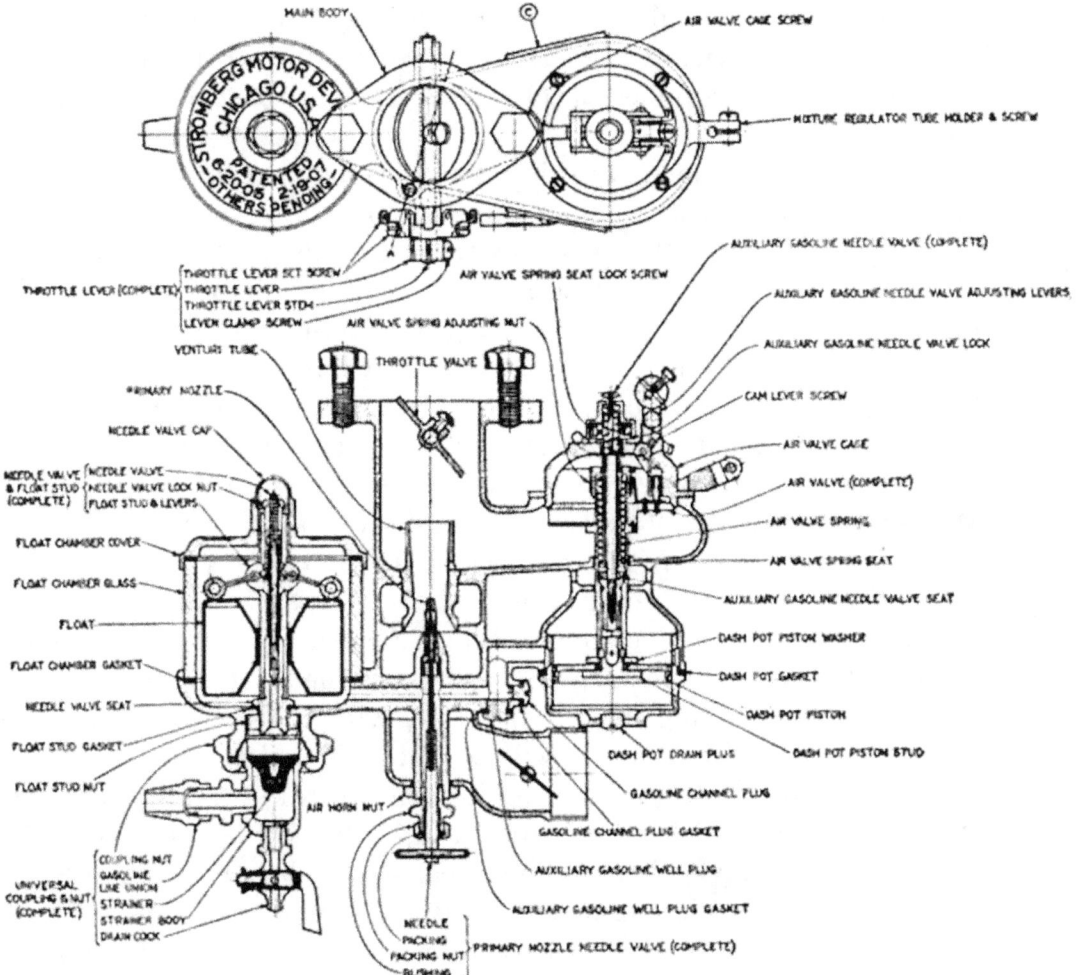

Fig. 109.—Stromberg Model H carburetor.

If the motor runs too fast with throttle closed, turn the small set screw in throttle stop to the left. If the motor stops when the throttle is fully closed, turn the set screw to the right.

As the throttle opens, the hot air damper, which is connected to it by a link, gradually closes, the greatest amount of hot air passing through the jackets when the throttle is nearly closed. The position of the hot air damper at any time is indicated by the slot at the end of the damper shaft. By loosening the set screw in the damper lever, this can be set for any desired relation between the damper and the throttle. Ordinarily the hot air damper should be nearly horizontal when the throttle is closed.

63. Stromberg Model H.—The Stromberg Model H carburetor, shown in Fig. 109, is of the double-jet type with two adjustments, for high and low speed, both working on the gasoline supply.

The gasoline in the glass float chamber is regulated by the hollow metal float. The fuel for low speed is furnished by the spray nozzle in the venturi tube, through which the low speed air passes. The adjustment for this nozzle is by means of the needle valve, as shown.

At high speed, the auxiliary air comes through the auxiliary air valve, which in turn automatically regulates the gasoline flow from the auxiliary gasoline valve. This supplies the extra gasoline for high speed and heavy duty service.

The dash pot with the piston riding in gasoline prevents all fluttering of the air valve on its seat when opening and closing.

This type of carburetor is fitted with a strangling or choke valve in the primary air inlet, for starting in cold weather. This assists in the vaporization of the gasoline by increasing the suction on the liquid.

The spring tension on the air valve and auxiliary needle valve is controlled either from the dash or from the steering post, depending upon the style of control installed. This permits adjustments to be made in order to compensate for varying conditions of weather, fuel, and operation.

64. Zenith Model L.—This carburetor, shown in Fig. 110, differs from most conventional types in the absence of auxiliary air valves. It is a "fixed" adjustment carburetor, and has as its particular feature the "compound nozzle." The compound nozzle consists of an inner nozzle, the gasoline for which is furnished direct from the float chamber. The amount of gasoline leaving this nozzle varies with the suction and consequently the mixture from this nozzle would be too rich at high speeds. To compensate for this rich mixture, the compensating nozzle surrounding the main nozzle furnishes a mixture "too weak" at high speeds. This is because the gasoline feed to this jet is constructed so as to be constant at all speeds. When the engine speeds up, the amount of air increases and the compensating mixture is a weak one. This answers the purpose of the auxiliary air valve on other types of carburetors and keeps the mixture of constant proportions. By a proper selection of the two nozzles a well balanced mixture can be secured through the entire range.

In addition to the compound nozzle, the Zenith is equipped with a starting and idling well. This well terminates in a priming hole at the edge of the butterfly valve, where the suction is greatest when the valve is slightly open. The gasoline is drawn up by the suction at the priming hole and, mixed with the air rushing by the butterfly, gives a rich slow speed mixture. The slow speed mixture is regulated by the regulating screw, which admits air to the priming well. At higher speeds with the butterfly valve opened, the priming well ceases to operate and the compound nozzle drains the well and compensates for any engine speed.

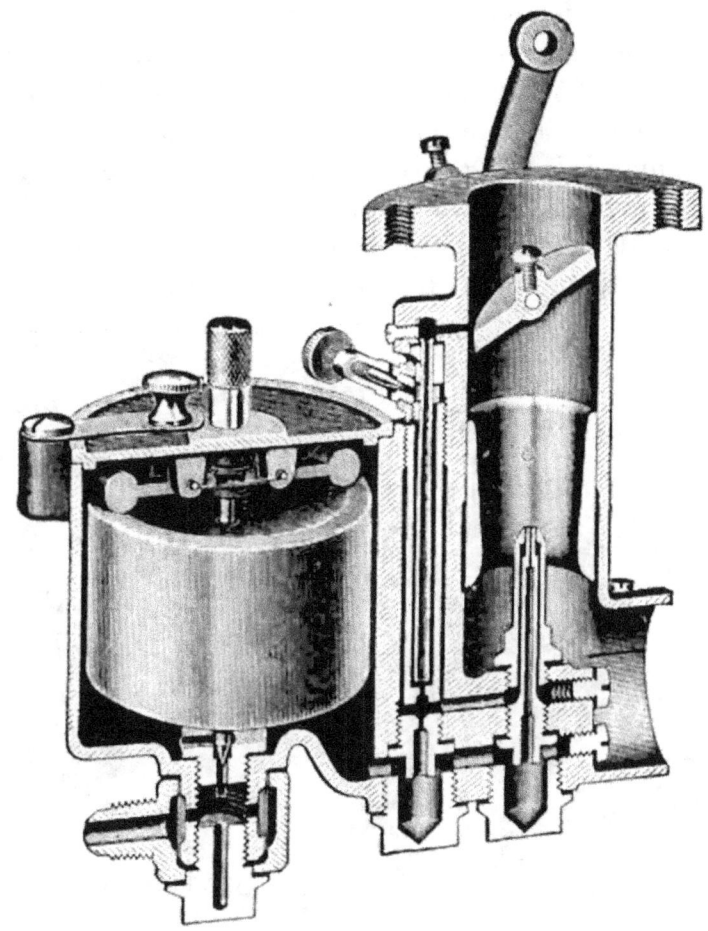

Fig. 110.—Zenith Model L carburetor.

65. Rayfield Model G.—This carburetor is illustrated in Figs. 111 and 112. It has two jets and the gasoline is drawn through them into the mixing chamber, the quantity being controlled by adjustments on the outside of the carburetor. As will be noticed, there are no air valve adjustments, but two gasoline adjustments, a low speed adjustment and a high speed adjustment. The names of the lettered parts on Fig. 111 are as follows:

D —Throttle Arm.	M —Regulating Cam.
G —Priming Lever.	S —Drain Cock.
H —Gas Arm.	U —Needle Valve Arm.
	X —Drain Cocks.

The suction created by the downward motion of the motor pistons draws air into the mixing chamber through the primary and auxiliary air inlets. This air rushes through the mixing chamber, around the nozzle and the metering pin, and picks up the gasoline which leaves the nozzle and jet in the form of a spray. Thus the action of the mixing chamber is not unlike that of an ordinary atomizer in which the air, forced from the rubber bulb, picks up a certain amount of the liquid in the bottle and sprays it out in the form of a fine vapor.

That the proportion of air and gasoline in the mixture may be correct for all motor speeds, one fixed air inlet and two variable auxiliary air inlets are provided. The lower air valve opens and closes with the main or upper automatic air valve, giving a greater volume of air in proportion to the greater amount of gasoline to be vaporized. In other words, at high motor speeds or when the throttle is fully opened, the motor requires more gas and consequently a greater volume of air to vaporize the gasoline

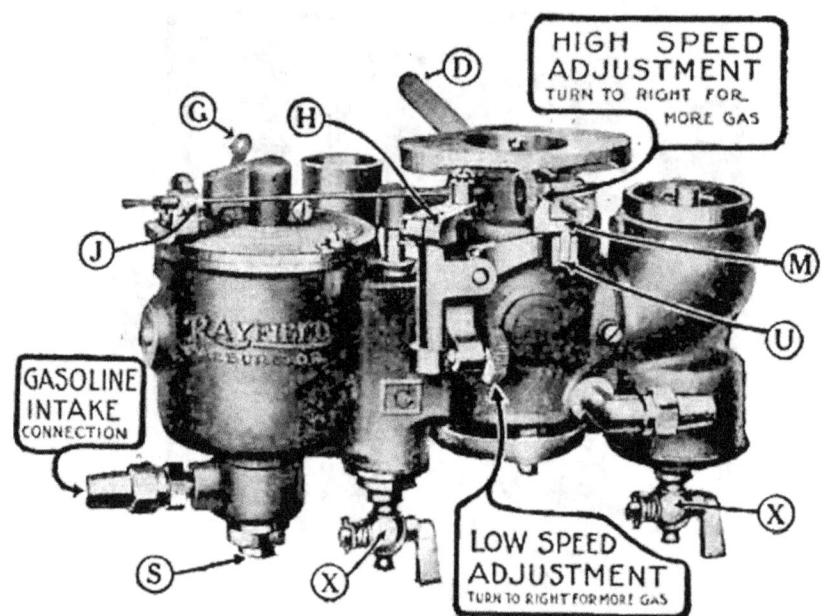

FIG. 111.—Rayfield Model G carburetor.

which comes through the spray nozzles; at low motor speeds, less gas is required and consequently less air is necessary to vaporize the gasoline.

At the front end of the carburetor is the main auxiliary air valve. This is controlled by a spring and dashpot. At low speeds, when only a small amount of air is being drawn through the carburetor, the spring and dashpot hold this valve almost shut. As the speed increases and more air is needed, the suction operating against the tension of the spring draws the valve further and further open, thus giving an increased supply of air in proportion to the need for the increased speed. The motion of this valve moves the metering pin and admits an additional supply of gasoline at this second nozzle.

Rules for Adjusting Rayfield Model G.—With throttle closed, and dash control down, close the nozzle needle by turning the low speed adjustment to the left until block U slightly leaves contact with the cam M. Then turn to the right about three complete turns. Open the throttle not more than one-quarter. Prime the carburetor by pulling steadily a few seconds on the priming lever G. Start the motor and allow it to run until warmed up. Then with retarded spark, close the throttle until the motor runs slowly without stopping. Now, with the motor thoroughly warm, make the final low speed adjustment by turning the low speed screw to the left until the motor slows down and then turn to the right a notch at a time until the motor idles smoothly.

To make the high speed adjustment, advance the spark about one-quarter. Open the throttle rather quickly. Should the motor back-fire, it indicates a lean mixture. Correct this by turning the high speed adjusting screw to the right about one notch at a time, until the throttle can be opened quickly without back-firing.

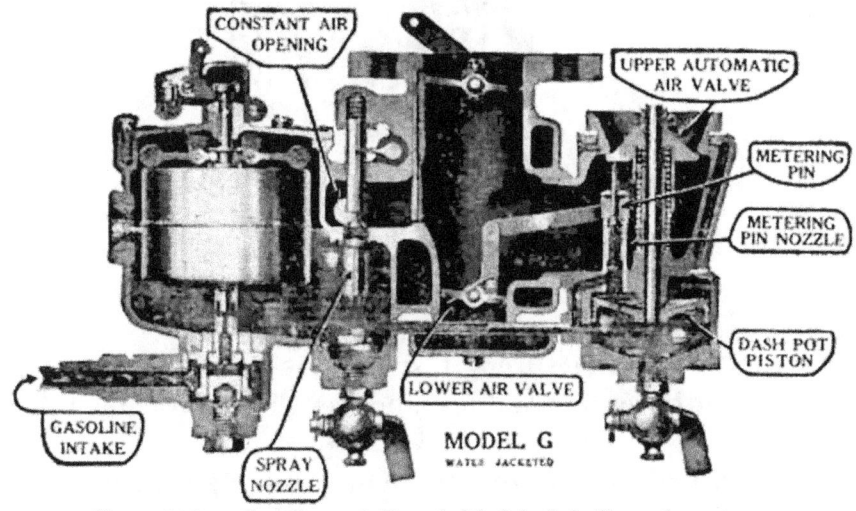

Fig. 112.—Section of Rayfield Model G carburetor.

If "loading" (choking) is experienced when running under heavy load with throttle wide open, it indicates too rich a mixture. This can be overcome by turning the high speed adjustment to the left.

66. Carter Model C.—The Carter carburetor, shown in section in Fig. 113, is of unconventional design and construction in many ways. The float is of copper and is spherical in shape. The float valve is provided with a shock absorber to prevent the valve from pounding on its seat when the car is being driven over rough roads.

There are three adjustments, for low, intermediate, and high speeds. The adjustable fuel tube gives the advantages of multiple jets. For low speeds the air taken in just above the bottom of the fuel tube takes gasoline from around the bottom of the tube. Under increased suction the gasoline is sucked higher in the tube and is sprayed through a number of openings in the side of the fuel tube into the air coming through the intermediate air valve. The high speed air adjustment is made from a lever connection on the dash.

67. General Rules for Carburetor Adjustment.—Very few general rules can be given for the adjustment of a carburetor. It is usually a very wise plan to let well enough alone, but if adjustments are necessary, it is very essential that they be made by someone familiar with the carbu-

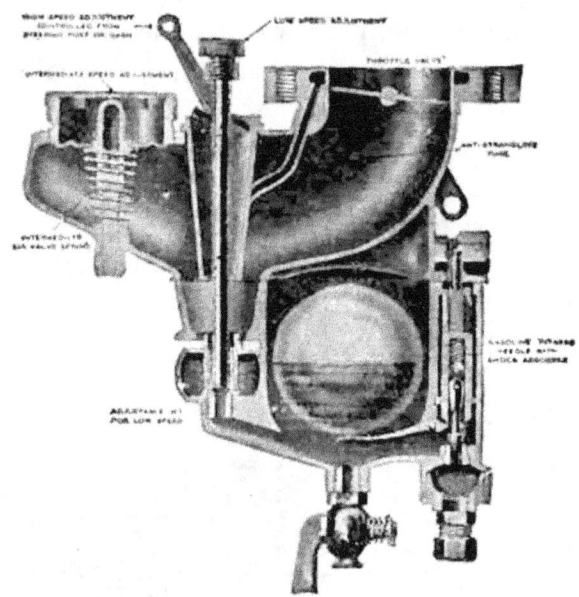

Fig. 113.—Carter Model C carburetor.

retor, or that the manufacturers' instructions be followed out in detail. The common carburetor troubles and remedies will be taken up in Chap. IX.

On most types of carburetors, there are two adjustments to be made, a low speed adjustment and a high speed adjustment. The low speed adjustment is made with the engine running idle, the spark retarded, and the throttle about one-quarter open. This is usually the gasoline adjustment. The high speed, or auxiliary air adjustment, is made with the engine running with throttle open and spark advanced. In all cases the adjustment should be made after the engine has warmed up to its normal running temperature.

Judging the mixture is largely a matter of experience. A rich mixture is indicated by the overheating of the cylinders, waste of fuel, choking of the engine and mis-firing at low speeds, and by a heavy black exhaust smoke with a very disagreeable odor. A weak mixture manifests itself by back-firing through the carburetor and by loss of power. A proper mixture will give little or no smoke at the exhaust. Blue smoke is caused by the burning of excess lubricating oil and has no relation to the quality of the mixture.

68. Carburetor Control Methods.—The carburetor is controlled from the driver's seat. The hand throttle on the steering post regulates the amount of mixture to the cylinders, thus regulating the engine and car speed. In conjunction with the throttle connection, is the accelerator on the toe-board, which permits the throttle to be opened by the foot, independently of the hand lever. The accelerator must be held open by the pressure of the foot. As soon as pressure is removed from it, the throttle closes to the point set by the hand lever. The air and gasoline adjustments are usually made from the dash of the car.

69. The Gravity Feed System.—There are numerous systems for feeding the gasoline to the carburetor from the gasoline tank, which may be placed at the rear of the frame, in the cowl, or under the seat. These feed systems are classified as gravity, pressure, and vacuum systems.

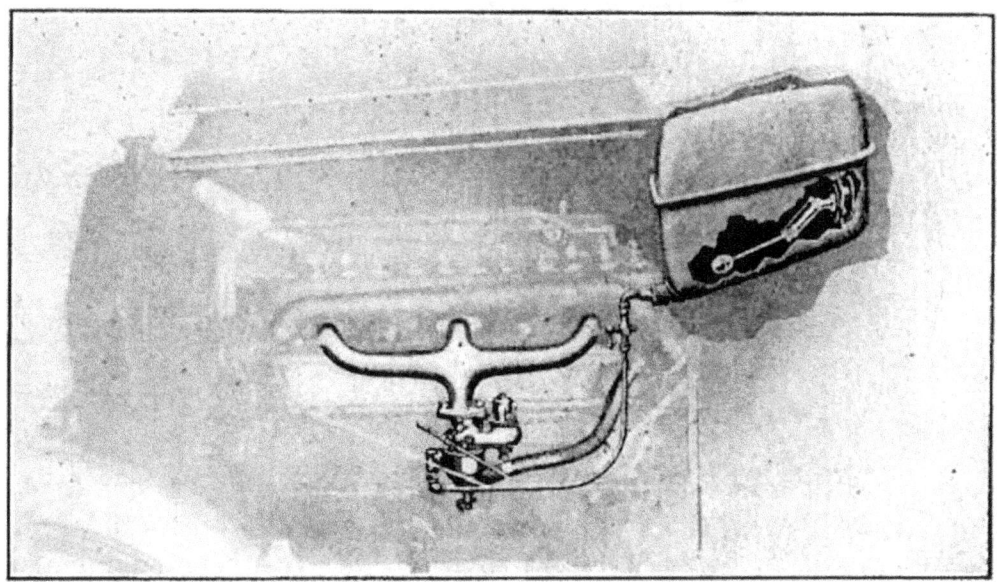

Fig. 114.—Studebaker gravity feed system.

In the gravity system of gasoline feed, the fuel flows to the carburetor by gravity alone. The tank may be placed either under the seat or in the cowl. If under the seat, there is the disadvantage of having to remove the cushions before being able to fill the tank. There is also the possibility in some cases that the tank will become lower than the carburetor when going up hill, and consequently the gasoline will not flow. Both of these disadvantages are done away with by placing the tank in the cowl. In either case, however, the pressure on the carburetor float valve varies as the level in the tank varies. When filling the tank, any gasoline which spills or leaks either falls around the seat, in the car, or on the engine. The advantage of the gravity system is that it is simple and always ready. Figure 114 shows the gravity system used on the Studebaker car, with the tank in the cowl. This shows the float operating the gasoline indicator.

70. The Pressure Feed System.—When the gasoline tank is placed at the rear of the frame, it is obviously impossible to use the gravity system. By putting a pressure in the gasoline tank, the gasoline may be forced by pressure to the carburetor. The pressure is maintained by a small air pump operated by the engine, or by a hand pump, or both. After filling the tank, a hand pump is used to get up pressure until the engine has been started. A safety valve in the pressure system keeps the pressure from getting too high. A particular advantage of this type of feed

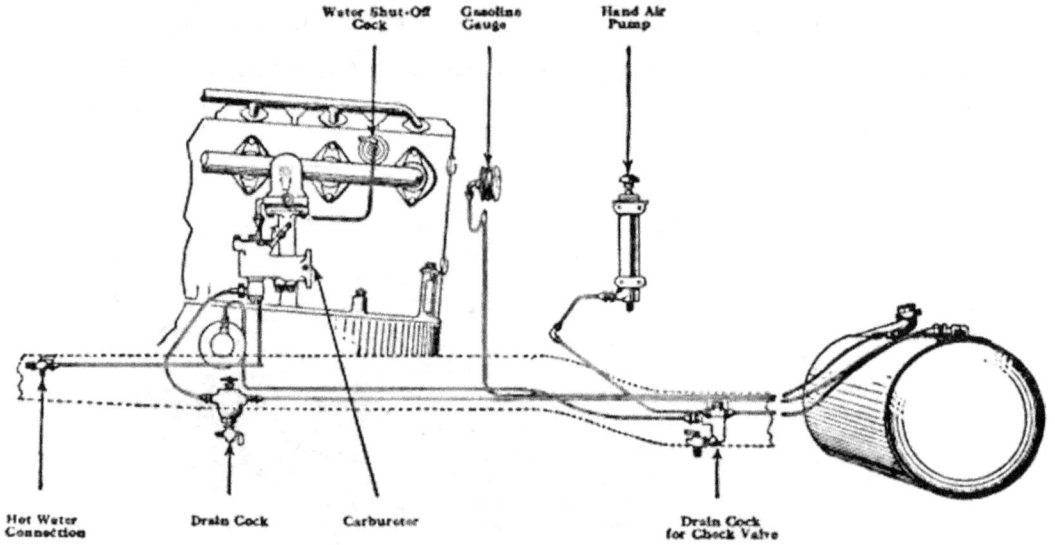

Fig. 115.—Pressure feed system.

system is that gasoline feeds to the carburetor regardless of the position of the car. As in the gravity system, the pressure on the float valve is liable to vary. The filler cap is placed away from the engine and passengers, and gasoline may be put in without disturbance. A typical pressure feed system is illustrated in Fig. 115.

71. The Vacuum Feed System.—Several systems have been developed in which the gasoline is transferred from the main tank at the rear of the car by vacuum, or suction, to a small auxiliary tank near the engine. From this small tank it flows by gravity to the carburetor. Figures 116 and 117 show the installation of the Stewart vacuum system in a car, and Fig. 118 indicates the construction of the auxiliary vacuum tank.

This system comprises a small round tank, mounted on the engine side of dash. This tank is divided into two chambers, upper and lower. The upper chamber is connected to the intake manifold, while another pipe connects it with the main gasoline tank. The lower chamber is connected with the carburetor.

The intake strokes of the motor create a vacuum in the upper chamber of the tank, and this vacuum draws gasoline from the supply tank. As the gasoline flows into this upper chamber, it raises a float valve. When this float valve reaches a certain height, it automatically shuts off the vacuum valve and opens an atmospheric valve, which lets the gasoline flow down into the lower chamber. The float in the upper

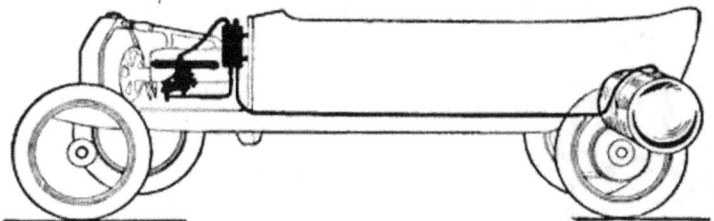

Fig. 116.—The Stewart vacuum feed system.

chamber drops as the gasoline flows out, and when it reaches a certain point it in turn reopens the vacuum valve, and the process of refilling the upper chamber begins again. The same processes are repeated continuously and automatically. The lower chamber is always open to the atmosphere, so that the gasoline always flows to the carburetor as required and with an even pressure.

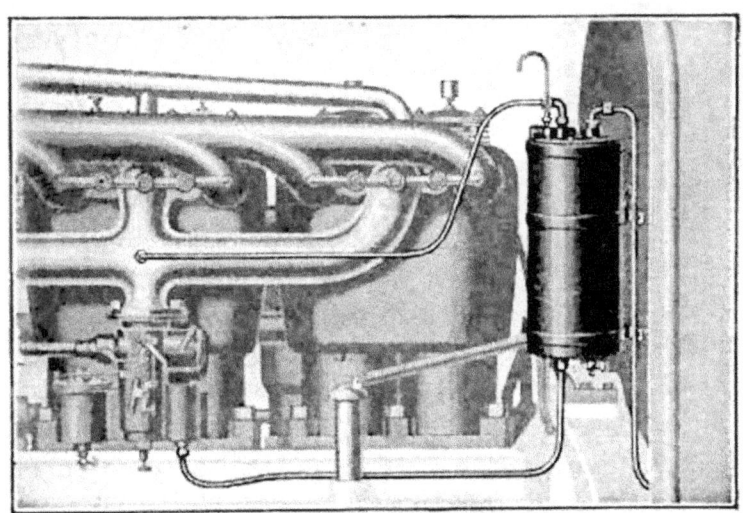

Fig. 117.—Under the hood.—The Stewart vacuum feed system.

The amount of gasoline always remaining in the tank gets some heat from the motor and thereby aids carburetion; it also makes starting easier, by reason of supplying warm gasoline to the carburetor. The lower chamber of the tank is constructed as a filter, and prevents any water or sediment that may be in the gasoline from passing into the carburetor. A petcock, in the bottom of the tank, permits drawing off this sediment and also allows the drawing of gasoline, if required for priming or cleaning purposes.

72. Intake Manifolds.—The tendency in present engine design is to make the intake manifold of such shape and proportions that the path from the carburetor to the engine cylinders shall be as short and smooth as possible. Being close to the cylinders, the manifold as well as the carburetor is heated, greatly aiding the vaporization of the gasoline. The short manifold gives the gas very little chance to condense between the carburetor and the cylinders. It is also desirable to have the distance from the carburetor to the different cylinders the same in all cases. This insures the same amount of mixture to each cylinder.

73. Care of Gasoline.—Gasoline, being a volatile liquid, is very dangerous if not properly handled, but if proper care and attention are given to it there should be no danger whatever. It should never be exposed in a closed room, as it will evaporate, mix with the air, and form a very explosive mixture. Open lights should always be kept away from gasoline in all cases. When it is necessary to handle gasoline at night, it should be done with an electric light. *Do not under any conditions use an open light.*

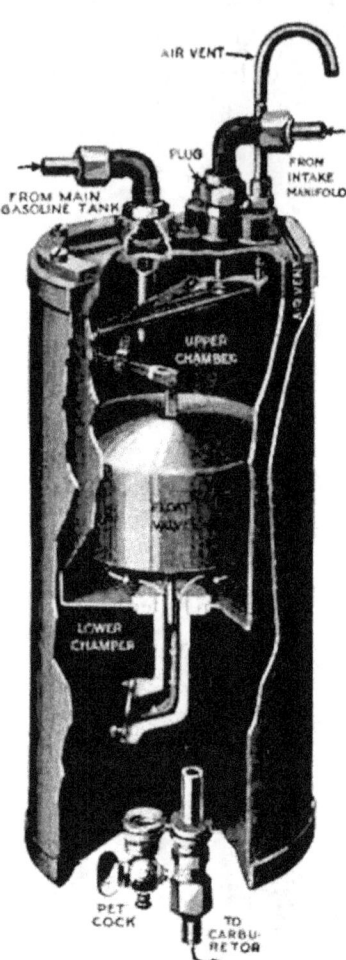

FIG. 118.—Stewart vacuum tank.

In putting out a gasoline fire, water will only spread the fire, as the gasoline, being lighter than water, floats on it. The only successful method of extinguishing a gasoline fire is to smother it, either by sand, or a blanket, or by the gases from a fire extinguisher.

The exhaust gases from a gasoline engine are very deadly. Do not breathe them for any length of time. If it becomes necessary to run your engine in a small garage with the doors closed, arrangement should be made to pipe the exhaust to the outside air.

CHAPTER V

LUBRICATION AND COOLING

74. Friction and Lubricants.—The purpose of lubrication is to reduce friction between moving surfaces. If parts moving on each other were not separated by a film of lubricant, the surfaces would rapidly rub away. Friction is a force that tends to retard the motion of one surface over another. The frictional force depends on the nature of the surface, and also on the kind of material. It is caused by the small projecting particles which extend from the surface. The rougher the surface and the softer the material, the greater the friction; or, the harder the material and the smoother the surface, the less the friction. The more friction there is, the greater the loss of power, as it requires power to overcome friction. A great amount of friction is necessary in certain parts of the car in order that they be efficient, such as in the brakes, the clutch, and the outer surface of the tires. On the other hand, it is essential that all friction possible be eliminated from the bearings in order to have as little of the motive power lost as possible.

The principal lubricants used are fluid oils, semi-solids, and sometimes solids, such as graphite. There are three general sources of lubricants: animal oils, such as lard, fish oil, etc.; vegetable oil, such as olive oil, linseed oil, etc.; and mineral oils, which are secured from petroleum. These lubricating mediums should each be used where they are best adapted. An oil that is suitable for one part of the mechanism may not be suited for another part. Only mineral oils should be used in gasoline engine cylinders, as they alone meet the requirements. For this reason the oils used for steam engine cylinders are not good for gasoline engine use, as they do not withstand the high temperature which rises in the gas engine cylinder. There are two main requirements for good cylinder oil. It should have a high flash point, that is, it should not break down and give off inflammable gases at low temperatures; and, second, it should retain its body and not become so thin as to be worthless as a lubricant at high temperatures. It should have sufficient body to maintain a positive film between piston and cylinder, yet should not be so heavy as to retard the free motion of the piston and rings. It should also be free from acids or any form of vegetable or animal matter. The vegetable or animal matter will decompose at high temperatures and gum up the cylinder. The acid will etch the smooth surface of the cylinder and cause excess friction. A simple method to test for acid is to dissolve a little of the oil in warm alcohol and then dip a piece of blue litmus paper in the solution. If there is any acid present, the paper will turn red. The litmus paper can be obtained at any drug store.

75. Cylinder Oils.—Cylinder oils are usually classified in three grades; light, medium, and heavy. Light cylinder oil looks something like the ordinary machine oil, and is slightly more viscous. The medium is somewhat heavier than the light, and might be compared to warm maple syrup. Light and medium oils should be used only on engines which have close-fitting pistons. The heavy oil is used in air-cooled engines and in engines that have loose pistons or that become too hot to use the lighter grade of oil. A good gas engine oil should have a high degree of viscosity at 100°F., a flash point not under 400°, and a fire test of over 500°.

76. Viscosity.—Viscosity is the property of a liquid by which it has a tendency to resist flowing. Oils are tested for viscosity by being put in a container and allowed to flow through a small opening. The oil that flows the fastest has the least viscosity. In some parts of the automobile it is necessary to use oil with less viscosity than in other parts. Tight fitting bearings should use oil with very little viscosity, while meshed gears should have semi-solid lubricants because the pressure on the rubbing surfaces is very high.

77. Flash Point.—The flash point is the temperature at which, if an oil be heated and a flame held over the surface, the vapor rising from the oil will burst into flame, but will not continue to burn. A thermometer is placed in the oil bath and the temperature taken at this point.

78. Fire Test and Cold Test.—Fire test is merely a continuation of the flash point test; that is, the temperature at which the vapor which rises from the oil will *continue* burning, and not merely flash for a second. Both these tests are used only on cylinder oil.

There is another test that is called the "cold test," which indicates the temperature at which the oil hardens, or becomes so stiff as not to flow. Good cylinder oil should not become so stiff as to prevent reaching the desired points at zero temperature.

79. General Notes on Lubrication.—There is no one thing which is the primary cause of more trouble and the cause of more expense in maintenance to the mechanism of an automobile than insufficient lubrication.

All moving parts of a car are usually manufactured with a high degree of accuracy and the parts are carefully assembled. In order to maintain the running qualities of the car it becomes necessary to introduce systematically suitable lubricants between all surfaces which move in contact with one another.

The special object of this chapter is to point out the places in the car which require oiling. While it is manifestly impossible to give exact instructions in every instance as to just how frequently each individual point should be oiled or exactly how much lubricant should be applied, we can give this approximately, based on average use.

It should be borne in mind that friction is created wherever one part moves upon or in contact with another. Friction means wear, and the wear will be of the metal itself unless there is oil, and oil is much cheaper than metal. The use of too much oil is better than too little, but just enough is best.

Proper lubrication not only largely prevents the wearing of the parts, but it makes the car run more easily, consequently with less expense for fuel and makes its operation easier in every way.

The oiling charts shown in this chapter indicate the more important points which require attention. But do not stop at these. Notice the numerous little places where there are moving parts, such as the yokes on the ends of various connecting rods, and pull rods, etc. A few drops of oil on these occasionally will make them work more smoothly.

Oil holes sometimes become stopped up with dirt or grease. When they do, clean them out and be careful not to overlook them. Also be careful not to allow dirt or grit to get into any bearings.

Judicious lubrication is one of the greatest essentials to the satisfactory running and the long life of the motor car. Therefore lubricate, and lubricate judiciously.

The auto engine should be lubricated by some means that will insure a definite supply of lubricant to the moving parts and that will supply the loss caused from vaporizing, burning and leakage.

The differential, axle bearings and shift gears are lubricated with semi-solid grease. The rear axle is not oil-tight, and therefore a fluid oil should not be used. Semi-solid lubricants also help to cut down the noise and wear where the pressure is heavy, and have sufficient cushion so that they adhere to the gear teeth. The lighter oils are better adapted for the high speed close-fitting parts. Other moving parts may be lubricated with the ordinary oil can, but are generally lubricated by the compression cup system. These cups may be screwed up from time to time to add more lubricant to the bearing surfaces.

The transmission should always contain sufficient lubrication to bring it up to the level of the drain plug on the side of the case, or so that the under teeth of the smallest gear will enter to their full depth.

The differential case should contain enough lubricant to bring it up to the filling hole, or should be about one-third full.

Wheel bearings should be packed with a thin cup grease. Do not use a heavy grease because it will work away from the path of the roller or ball and will not return. In each hub there is usually a small oil hole. Inject some engine oil here whenever you are oiling the car. It will keep the grease soft and in good condition. Before lubricating any part, wipe all dirt from it so that the dirt will not get into the bearings.

The steering gear is perhaps one of the most important parts of the car to keep properly lubricated. Failure of the steering apparatus is a dangerous thing and a few drops of oil given to the oil cups and the various steering connections constitute a cheap and safe means of avoiding accidents. Most types of steering apparatus are packed with grease which, having no outlet, will remain. However, the grease will become dry and a little oil should be added from time to time.

Few motorists think of lubricating their brake connections. Mud and water will find their way into the brake mechanism and a squeeze of the oil can and a turn of the grease cups given daily will keep them in good working condition.

The principal engine lubricating systems can be grouped under the following heads: first, splash system; second, splash with circulating pump, which may be either a "forced feed" or a "pump-over" system; third, full forced feed; fourth, mixing the oil with the gasoline.

80. Splash System of Engine Lubrication.—The splash system is used in the Ford engine, as shown in Fig. 119. The oil is poured directly into the crank case until it comes above the lower oil cock. The level of the oil should be maintained somewhere between the two oil cocks. The flywheel runs in the oil and picks up some of it and throws it off by centrifugal force; some of the oil is caught in a tube and carried to the front end of the crank case where it lubricates the timing gears. As the oil flows back to the rear part of the crank case, it fills the small wells in the crank case under each connecting rod. As the connecting rod comes around, a small spoon or dipper on the bottom scoops up the oil, so that there is a regular shower of oil all the time. The pistons, cylinder walls, and bearings are lubricated in this manner and the oil is kept in continuous circulation. All parts of the clutch and transmission are lubricated in the same manner as the engine.

The oil level should never get below the lower oil cock and should never get above the upper oil cock. Never test the level of the oil when the engine is running.

81. Splash System with Circulating Pump.—This system has an oil reservoir or sump below the main crank case bottom. The oil from the sump in the lower half of the crank case is sucked through a strainer into the pump, usually at the rear end of the reservoir. The oil pump of the Buick engine is shown in Fig. 120. This pumps the oil up through a pipe to a sight feed on the dash so that the circulation can be observed by the driver. From here the oil returns to the splash trays in the lower

LUBRICATION AND COOLING

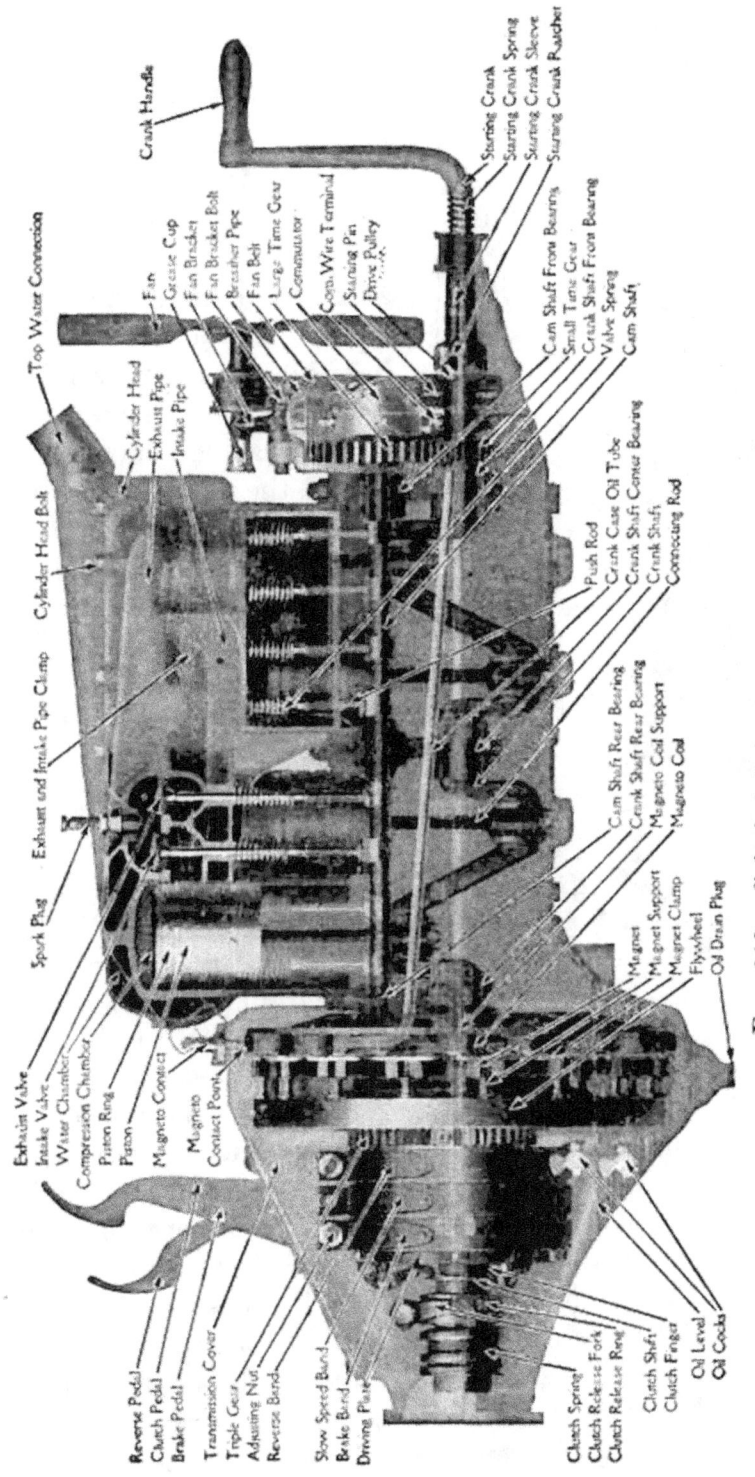

Fig. 119.—Splash lubricating system of Ford engine.

107

half of the crank-case through the distributor pipe. As the crank comes around, the spoons or dippers on the connecting rods dip into these trays and force some of the oil up into the crank pin bearings and splash the remainder over the interior of the crank case and up into the cylinders and pistons. As the oil drains back, it is caught in ducts and led to all the bearings of the motor, the excess running back into the sump to be used again.

The oil circulating pump consists of two small gears enclosed in a close fitting housing attached to the lower half of the crank case and driven by a vertical shaft and spiral gears from the cam shaft. As the gears turn, they take the oil into the spaces between the teeth and carry it around to the outlet where the action of the teeth meshing together squeezes the oil out of the spaces and forces it to flow to the sight feed on the dash. The pump requires no attention or adjustment except the addition of fresh oil to the crank case reservoir as often as is necessary to keep the oil level up to the oil cock. The sight feed on the dash merely shows whether or not the oil is circulating and does not show when the supply in the crank case is running low. Test the oil level at frequent intervals by opening the oil cock and see that the oil is kept up to this level. To remove the pump, draw off all the oil and take the pump out from below.

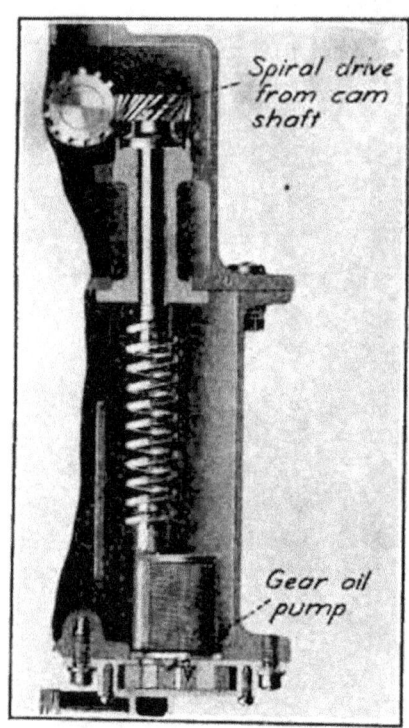

Fig. 120.—Buick oil pump.

The motor lubrication on the Overland car is shown in Fig. 121, and is the splash and pump-over system. The oil reservoir is located in the bottom of the crank case and is filled through the combination breather pipe and oil filler on the right side of the engine. The glass gauge on the side of the crank case close to the breather pipe indicates the oil level. The oil pump, which is located in the rear of the crank case, is driven from the cam shaft. The lubricant is drawn from the base and, after passing through a strainer, runs through a sight feed on the dash, and from there it runs into the troughs and is splashed into the bearing surfaces. It is very important that the oil strainer be kept clean at all times so that proper circulation of the oil is insured. For this reason the removal of the oil strainer has been made easy. By unscrewing the large plug on the side of the crank case right opposite the oil pump, the

LUBRICATION AND COOLING

cylindrical screen may be drawn out and cleaned by dipping into a pail of gasoline. The owner should see that the oil screen is cleaned every 200 miles of the first 1000 miles and after that every 500 miles.

The lubricant circulates freely through the system as long as the small wheel in the dash sight-feed revolves. But as soon as the wheel stops or the sight-feed glass shows clear, this is an indication that the oil supply is exhausted, or that there is an obstruction in the circulation of the oil which should be located and remedied immediately, since serious and expensive trouble will result from running the motor with an insufficient supply of oil.

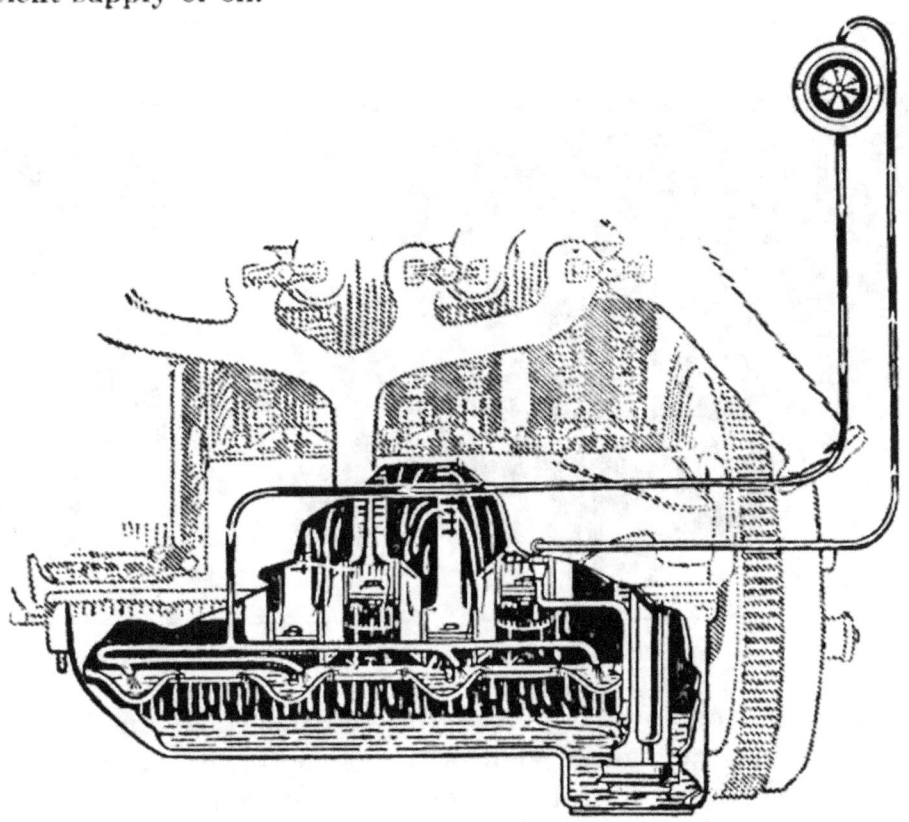

Fig. 121.—Overland splash system with circulating pump.

The wrist pin is lubricated from the cylinder walls, through the opening in the piston through which the wrist pin is inserted, as well as through a slot cut into the connecting rod over the wrist pin bushing.

The lubrication system of the Studebaker Four, Fig. 122, is called the constant level splash system combined with a forced feed to the timing gears. A quantity of oil is carried in a reservoir F, which is formed by the crank case of the motor. A pump B of the plunger type draws the oil from this reservoir and sprays it (G) over the connecting rod bearings. It also pumps surplus oil through a sight feed J or indicator on the dash, from which it flows over the timing gears D at the front of the motor and returns to the reservoir through the pipe U. The oil draining from the spray collects in troughs E which maintain a constant level of oil just under the connecting rods. At each revolution short projections M from the connecting rods dip into these troughs and splash oil over the lower ends of the pistons, and over the cam and crank shaft bearings.

To fill the oil reservoir of the motor, pour the oil in through a funnel shaped tube H, which you find on the left side of the motor. This funnel shaped tube is called the "breather pipe." At the side of the "breather pipe" there is a gauge I which shows the amount of oil in the

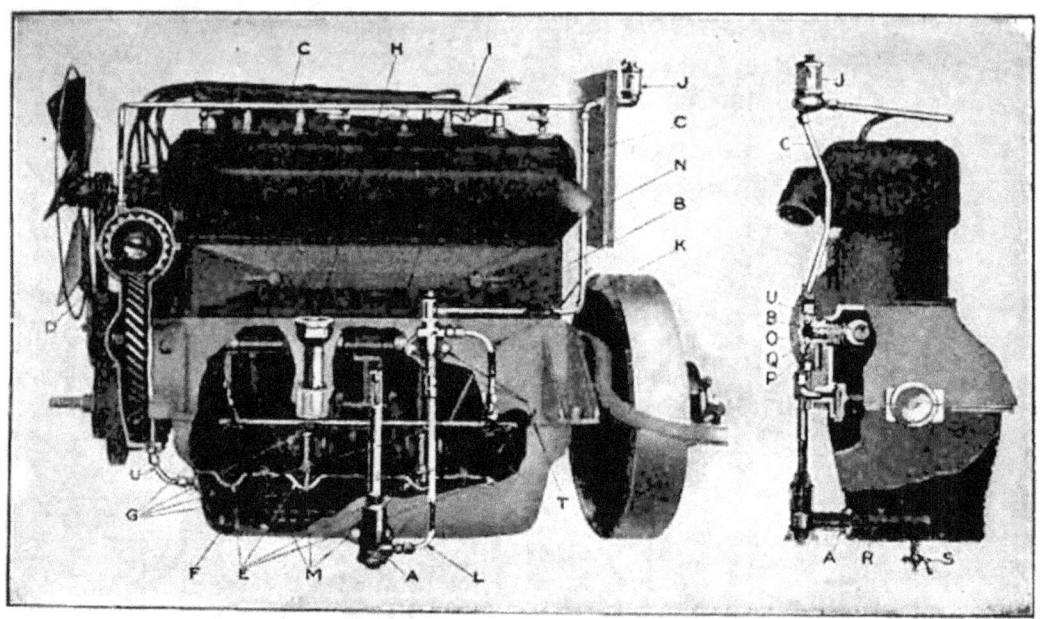

Fig. 122.—Studebaker splash system with forced feed.

reservoir. The oil is poured into the breather pipe until the gauge indicator rises to the highest point of the gauge, being careful that there is no more oil poured into the motor than just enough to bring the indicator to the highest point shown on the gauge. The only attention necessary to keep the motor perfectly lubricated is to see that the gauge indicator shows that there is oil in the reservoir.

When the motor is running, oil drops through a glass indicator or "sight feed" J on the dash. This "sight feed" can be seen from the seat and should not be forgotten by the driver. If the oil should cease to flow through the "sight feed" when the motor is running, the motor should be stopped and hood lifted to ascertain if the gauge I shows oil in the reservoir. If it does show oil in the reservoir, then either the oil pump or the connecting oil pipes are clogged and should be cleaned out.

LUBRICATION AND COOLING

82. Full Forced Feed System.—A full forced feed as used on the Cadillac Eight is shown in Fig. 123. A gear pump located at the forward end of the motor and driven from the crank shaft takes the oil up from the oil pan in the lower part of the crank case and forces it through a reservoir pipe running along the inside of the crank case, from which pipe there are leads to each of the main bearings. The crank shaft and webs are drilled and oil is forced from these main bearings to the connecting rod bearings through the drilled holes. The forward and rear bearings supply the rod bearings nearest them, while the center bearing

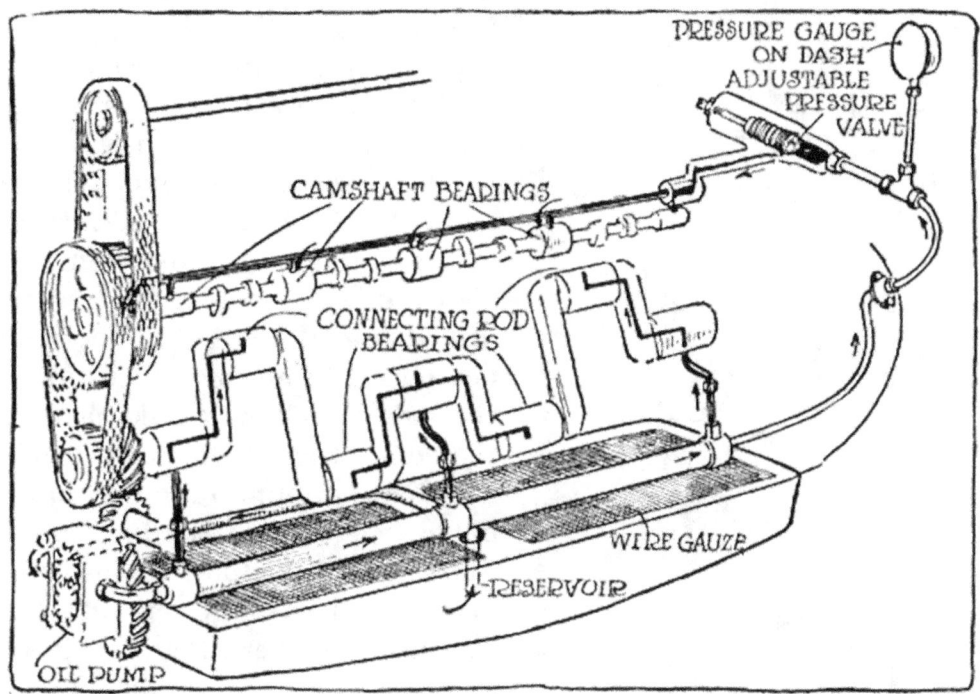

Fig. 123.—Cadillac forced feed oiling system.

takes care of the rod bearings on either side of it. The oil is then forced from the main reservoir pipe up to the relief valve, which maintains a uniform pressure above certain speeds, and overflows from this valve to a pipe extending parallel with the cam shaft and above it. Leads from this latter pipe carry lubricant by gravity to the cam shaft bearings and front end chains. Pistons, cylinders and piston pins get their oiling by the oil thrown from the lower ends of the connecting rods.

A gauge indicating the level of the oil is attached to the upper cover of the crank case. Whenever the indicator reaches the space marked "fill," oil should be added until the indicator returns to "full." A filling hole is provided in each block between the second and third cylinders. If the hand on the pressure gauge on the cowl vibrates or returns to zero on the dial when the engine is running, it indicates that the oil level is very low. Should this occur through neglect to add oil at the proper time, the engine should immediately be stopped and sufficient oil added to bring the pointer up to the top of the gauge before the engine is again started.

The hollow crank shaft oiling system as used by the Wisconsin Motor Mfg. Co. is shown in Fig. 124 and operates as follows:

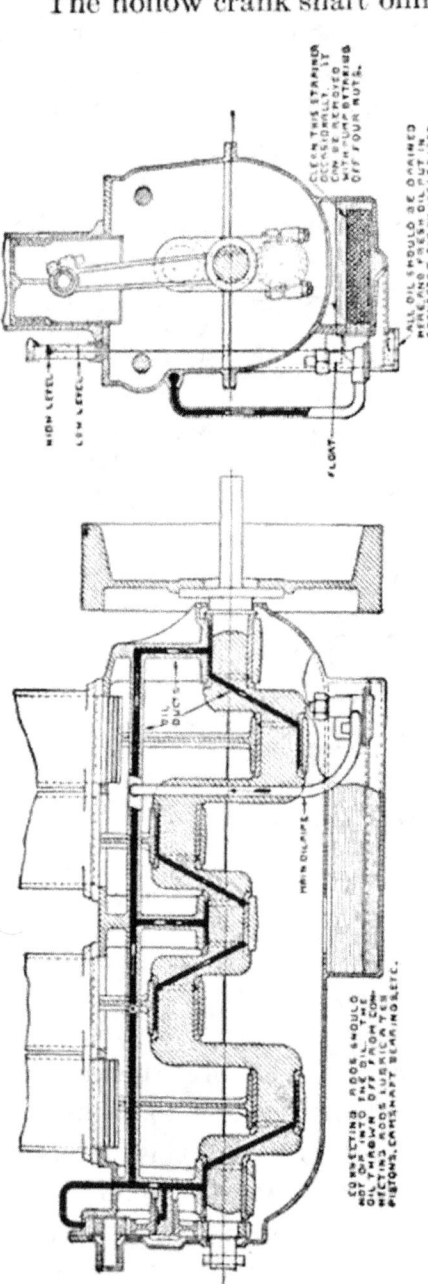

Fig. 124.—Forced feed oiling system used by Wisconsin Motor Mfg. Co.

The oil is carried in an independent chamber at the bottom of the crank case, and the connecting rods are not allowed to dip into this, thus preventing the oil from being whipped to a froth, and preserving its viscosity.

It is pumped by means of a gear pump located at the lowest point of the oil reservoir into a main duct, which is cast integral with the crank case, and from here distributed by means of ducts, drilled into the webs, to the main bearings. From here it is forced through a hollow crank shaft to the connecting rod bearings, and a sufficient amount of oil is forced out of the ends of the bearings to lubricate the pistons, piston pins, and cam shafts. A separate lead runs directly over the timing gears, and all oil is thoroughly filtered before it is pumped over again. An oil gauge indicates by means of a ball and float the exact amount of oil contained in the reservoir, and distinct marks on the glass gauge show the high and low mark, and if the oil is maintained between these two levels no burnt oil smoke will be emitted, and the spark plugs will not be fouled.

The pressure of the oil increases with the speed of the motor, so the faster the motor is run the more oil is forced to it, and *vice versa*. The location of the oil reservoir permits the proper cooling of the oil, thus minimizing the danger of burning out bearings.

The lubricating system for Knight sliding sleeve motors is also of the forced feed type. The following description is of the system used on the Moline-Knight car. Oil is drawn from the sump by a gear pump driven off the end of the eccentric shaft, and is delivered to the three main bearings, and the magneto drive shaft bearing under a pressure determined by the settings of a spring controlled by-pass valve, through which the excess oil is delivered. This excess oil is led to the chain driving the eccentric shaft and magneto, and flows thence to a trough and through a screen to the sump. Part of the oil delivered to the main bearings passes through holes in the crank shaft web to the crank pins, and thence through the tubular connecting rod to the hollow piston pins. From the two ends of the latter it flows to the sleeves and is distributed through holes and oil grooves in the latter over their circumference and the cylinder walls. All parts requiring lubrication not mentioned above are oiled by splash from the crank shaft and connecting rods. The flow of oil delivered under pressure is determined by a valve which is so connected as to open and close with the throttle. There are no oil grooves in any of the crank shaft bearings. The entire bottom of the crank case is covered by a screen, through which the oil returns to the sump.

83. Mixing the Oil with the Gasoline.—Another system that is used to some extent in two-stroke marine engines is to mix the lubricating oil with the gasoline, in the proportion of 1 pt. of oil to 5 gal. of gasoline. The easiest way is to thoroughly mix 1 pt. of oil with 1 gal. of gasoline, pour it into the fuel tank and then add 4 gal. of gasoline. The oil stays in solution with the gasoline. This system is very simple, as the lubricating becomes automatic and there are no regulators to adjust.

When the piston is on the up stroke, a charge of gasoline and oil is drawn through the carburetor. Here the oil and gasoline separate because the oil does not evaporate and the gasoline does. The gasoline mixes with the air in the form of a gas. The oil collects in the form of small globules which float in the mixture of gas and air and are carried into the crank case by the suction of the motor. Here some of the oil settles on the connecting rod and crank and flows through a special oil duct to the crank pin.

On the down stroke of the piston, the gas and oil are forced through the by-pass into the cylinder where the remainder of the oil is deposited on the cylinder walls. This operation is repeated every revolution of the engine, a new film of oil being supplied each time.

84. Selection of a Lubricant.—The proper lubrication of the motor car is more important than any other item in its care. Only the best high grade oils should be used to lubricate the engine. Some engines require lighter oils than others on account of the close-fitting pistons and rings. It is better to follow the instructions sent out by the manufacturers in regard to the kind of oil to use rather than for the motorist to make his choice or to be directed by an oil salesman. The different companies run extensive tests and find out in that way which oil is best suited for their type of engine. The only way to get the best lubricants is to pay the price. Money saved by cheap oils or grease may be more than lost in worn-out bearings or cylinders.

The multiple-disc type of clutch is the only one in which any lubrication should be used, and the oil here should be drained off about every 1000 miles, the clutch well cleaned out with kerosene, and then filled with light machine oil, the amount, of course, depending upon the capacity of the case. All clutches that use any kind of facing, such as asbestos, raybestos, or leather, should never be lubricated, as the oil decreases the friction and causes slipping. Clutch leathers will retain their life and softness better if given an occasional treatment of neatsfoot oil and then wiped dry.

The planetary transmission system in the Ford automobile is encased so as to revolve in an oil bath.

The differential housing and sliding gear transmissions and all other parts that use either heavy cylinder oil, transmission oil, or graphite grease, should be thoroughly cleaned every 1000 miles, or thereabouts, and well flushed out with kerosene in order to remove all sediment and metallic dust that may be in the old grease. All wheel bearings are of the ball or roller anti-friction type, and are packed with semi-fluid grease which should be renewed about every 1000 miles.

An excess of grease in the transmission or differential case will be shown by leaking at the joints, on account of the difficulty of keeping these members absolutely tight and still free to run. If there is too much grease in the differential case, it will run along the axle shaft and out over the oil guard, which is to prevent it from getting on the tire and also from interfering with the action of the internal brake.

Excess of lubrication in the engine will produce carbon deposits and dirty spark plugs. It may also cause the piston rings to gum up and stick. It can be detected by the color of the exhaust smoke, which will have a bluish tinge, or it may be detected by a sticky black coating on the spark plug.

A small amount of graphite and oil or grease should be supplied between the leaves of the springs. This can generally be done by jacking up the frame so that all weight is taken off the wheels, and by using a small clamping device with wedge-shaped jaws, which can be used to spread the leaves apart.

85. Directions for Lubrication.—A very good chart for lubrication purposes is sent out by the Chalmers Motor Car Co., and of course can be used for other standard makes of cars. This chart is as follows:

LUBRICATION AND COOLING

DIRECTIONS FOR LUBRICATION

EVERY DAY CAR IS IN USE, OR EVERY 100 MILES:

Part	Quantity	Lubricant
Crank case.	Keep oil at level of top try cock.	Motor oil.
Steering knuckle grease cups.	One complete turn.	Cup grease.
Steering cross rod grease cups.	One complete turn.	Cup grease.
All spring bolt grease cups.	Two complete turns.	Cup grease.
Speedometer driving gears.	One complete turn.	Cup grease.
Eccentric bushing of steering gear.	10 or 15 drops.	Motor oil.
Wheel hub oilers.	10 drops.	Motor oil.

TWICE A WEEK, OR ABOUT EVERY 200 MILES:

Part	Quantity	Lubricant
Fan hub bearing.	Few drops.	Motor oil.
Pump shaft grease cups.	Two complete turns.	Cup grease.
Steering gear case oiler.	Fill.	Motor oil.
Steering gear case grease cup.	Two complete turns.	Cup grease.
Steering wheel oil hole.	8 or 10 drops.	Motor oil.
Steering column.	10 or 15 drops.	Motor oil.

EVERY WEEK, OR ABOUT EVERY 300 MILES:

Part	Quantity	Lubricant
Spark and throttle shafts.	Few drops.	Motor oil.
Control bracket bearings.	Thoroughly.	Motor oil.
Transmission case.	Enough to cover lower shaft.	Motor oil.
Pedal fulcrum pin.	Thoroughly.	Motor oil.
Brake pull rods and connections.	Thoroughly.	Motor oil.
Brake cross rod grease cups.	Two complete turns.	Cup grease.
Torque rod grease cups, front and rear.	Two complete turns.	Cup grease.
Brake shafts on rear wheels.	Thoroughly.	Motor oil.
Rear spring perch grease cups.	Two complete turns.	Cup grease.

TWICE A MONTH, OR EVERY 500 MILES:

Part	Quantity	Lubricant
Magneto bearings (3 oil holes).	3 or 4 drops each.	High grade light machine oil.
Dynamo drive shaft universal joints.	Fill one-half full.	Cup grease.

EVERY MONTH, OR EVERY 1000 MILES:

Part	Quantity	Lubricant
Crank case.	Drain off dirty oil; clean oil screen at left of motor thoroughly; fill to level of top try cock.	Motor oil.
Reach rod boots.	Pack thoroughly.	Cup grease.
Spring leaves. (Jack up frame and pry leaves apart.)	Thoroughly.	Graphite grease.
Hub caps.	Pack thoroughly.	Cup grease.
Universal joints.	Remove grease hole plug and fill one-half full.	Cup grease.
Gasoline pressure hand pump.	4 or 5 drops on leather plunger.	Light machine oil.

EVERY 2000 MILES:

Part	Quantity	Lubricant
Differential housing.	3 pt.	Special axle compound.
Transmission case.	Drain thoroughly, flush with kerosene, refill to cover top lower shaft try cock.	Motor oil.

Dynamo should be lubricated every 3000 to 5000 miles.

When changing tires, put a few drops of oil on inside sliding ring of demountable rims to insure easy detaching.

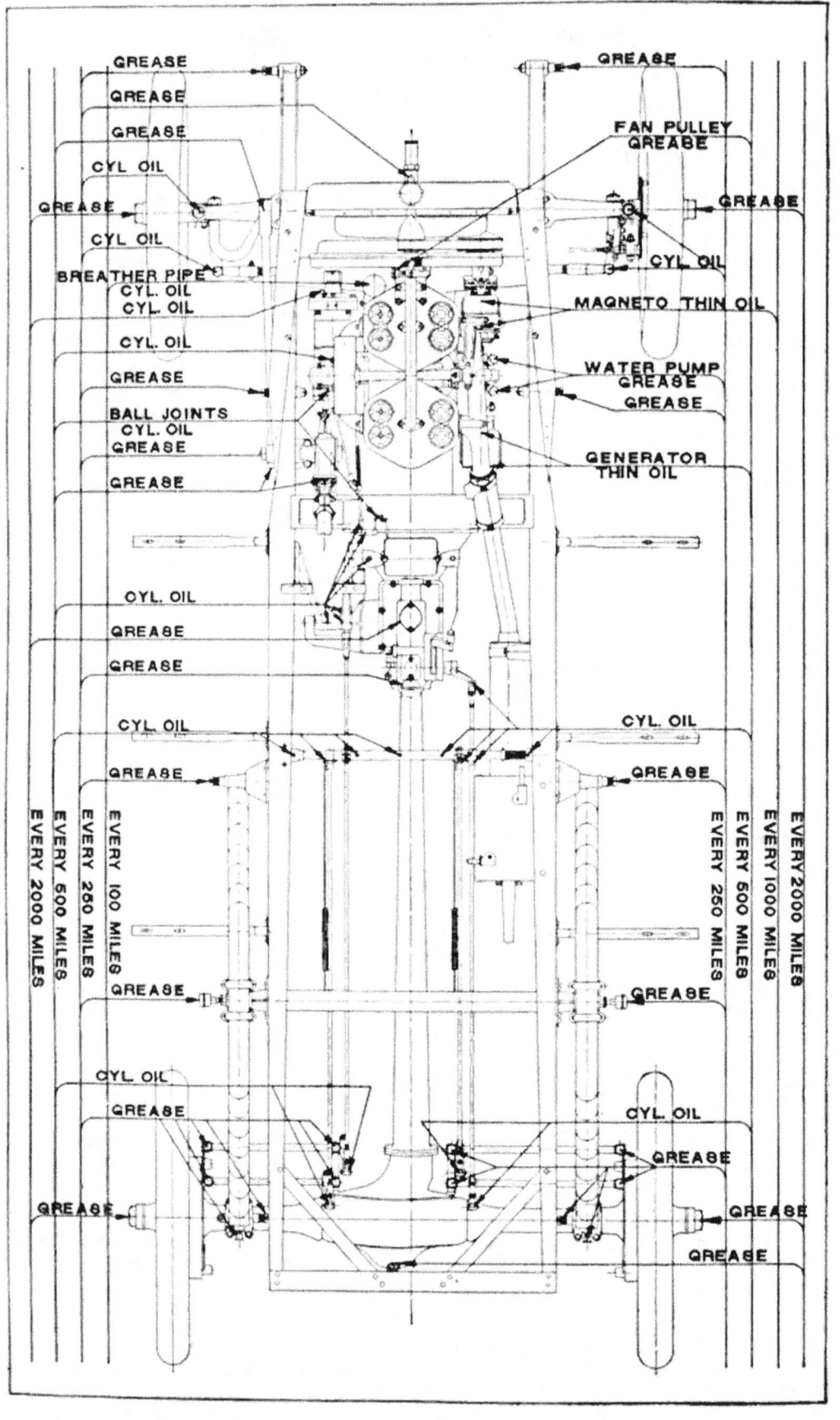

Figure 125 shows the location of the various places to be lubricated and the proper intervals for lubrication. This is the chart for the Case car.

86. Cylinder Cooling.—When an explosion occurs inside the cylinder of a gas engine, the gases on the inside reach a temperature of from 2000° to 3000°F. The walls of the cylinder are, of course, exposed to this high heat and would very quickly get red hot if we did not have some way of keeping them cool. The polished surface upon which the piston slides would be very quickly spoiled. The most common way of keeping a cylinder cool is by the use of water. Surrounding the cylinder is a metal jacket enclosing a space for the cooling water. By keeping a supply of water passing through this space, the cylinder can be kept cool enough for the operation of the engine. The cylinder head is also cast with a double wall, especially around the valves, so that these parts will also be kept cool. The cooling fluid used is generally water.

Water should not be allowed to remain in the jacket of an engine over night if there is danger of a frost, as the freezing of the water will crack the cylinder. When the supply of water is limited, as in an automobile, the water is cooled in a radiator or system of pipes, and then is used over again. The water is kept in circulation by a pump, or by the thermo-syphon system, and the hot water is cooled by the air passing over the radiator.

The circulation in the thermo-syphon system is based on the fact that cold water is heavier than hot water, and consequently, the water heated in the cylinder jackets flows up and over into the top part of the radiator, where it is cooled and then flows from the lower portion of the radiator back to the engine cylinder. Circulation is automatically maintained as long as the engine is hot and there is enough water in the radiator so that the return connection from the cylinder to the radiator contains water. This means that the radiator must be kept practically full all the time, or else there will be no circulation and the water will merely boil away.

When the pump system of circulation is used, the radiator may be lighter than in the syphon system, as less water is needed to do the same amount of cooling. The pump is driven from the engine, and the faster the motor runs the faster the water circulates. The centrifugal type of pump is generally used for circulating cooling water.

87. Water Cooling Systems.—Radiators differ in design. In some types the water flows through tubes of very small diameter. In this type it is necessary to have a circulating pump of some kind. In radiators having tubes of larger diameter, the thermo-syphon system may be used. The radiators using the small pipes have a greater capacity for their size because they have more exposed area for cooling in comparison with the amount of water they carry. The small tubes have the disadvantage of increased resistance. This is why it is necessary to use a pump.

The air for cooling purposes is usually drawn through the radiator by a fan placed directly back of it. This fan may be driven with a bevel or spur gear, with a silent chain, or with a wire or leather belt. In some cases, however, the engines are air-cooled, the cylinders being cast with a large number of fins or rings on the outer surfaces to increase the cooling effect of the air. In this case there is no water jacket.

The cooling system of the Overland is the thermo-syphon system, which eliminates the circulation pump and its gears, glands, stuffing boxes,

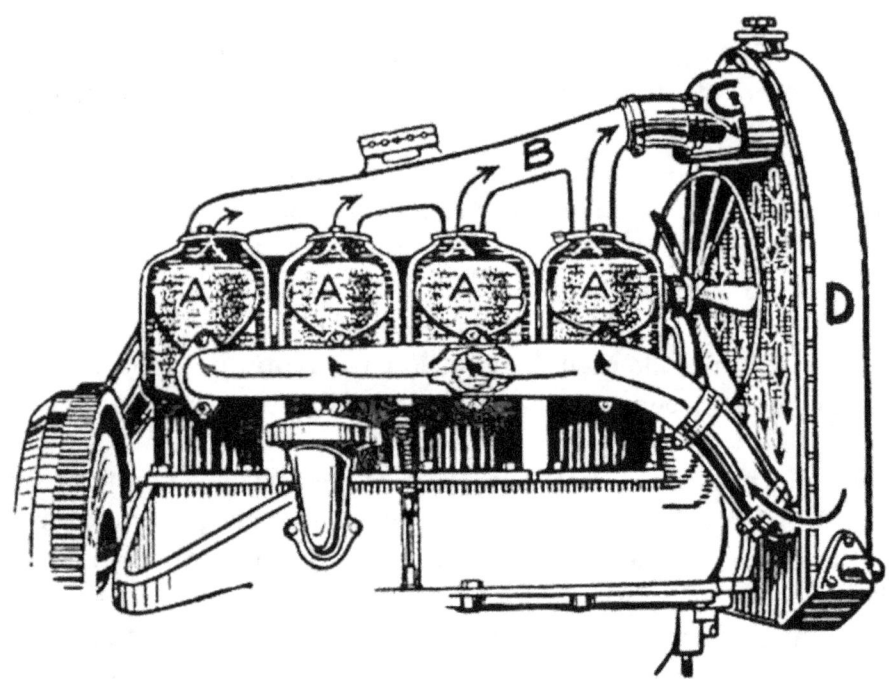

Fig. 126.—Overland thermo-syphon cooling system.

etc. The thermo-syphon system is automatic, as the speed with which the cooling water circulates is increased or decreased with every increase or decrease in jacket temperature. The action of the system is, briefly, as follows: The water enters the cylinder jackets A, Fig. 126. Upon becoming heated by the explosions within the cylinders, the water expands and, being lighter, rises to the top. It then enters the pipe B and passes into the radiator at C, where it is brought into contact with a large cooling surface, D, in the shape of the cellular radiator. On being cooled, and thereby contracting and becoming heavier, the water sinks again to the bottom of the cooling system, to enter the cylinders once more and to repeat its circulation. The cooling action is further increased by a belt-driven fan which draws air through the radiator spaces.

LUBRICATION AND COOLING 119

Figure 127 shows the cooling system on the Ford. This is also a thermo-syphon system, the principle of operation being the same as on the Overland. The arrows indicate the path of the cooling water.

The cooling system used on the Studebaker Four is the pump system shown in Fig. 128. The water system, which contains 10 qt. of water, consists of a radiator, hose connections, water line, pump, and water jackets which are incorporated with the cylinders. The radiator D being filled with water and the motor running, the centrifugal pump C forces the water to circulate as follows: From the pump it is driven

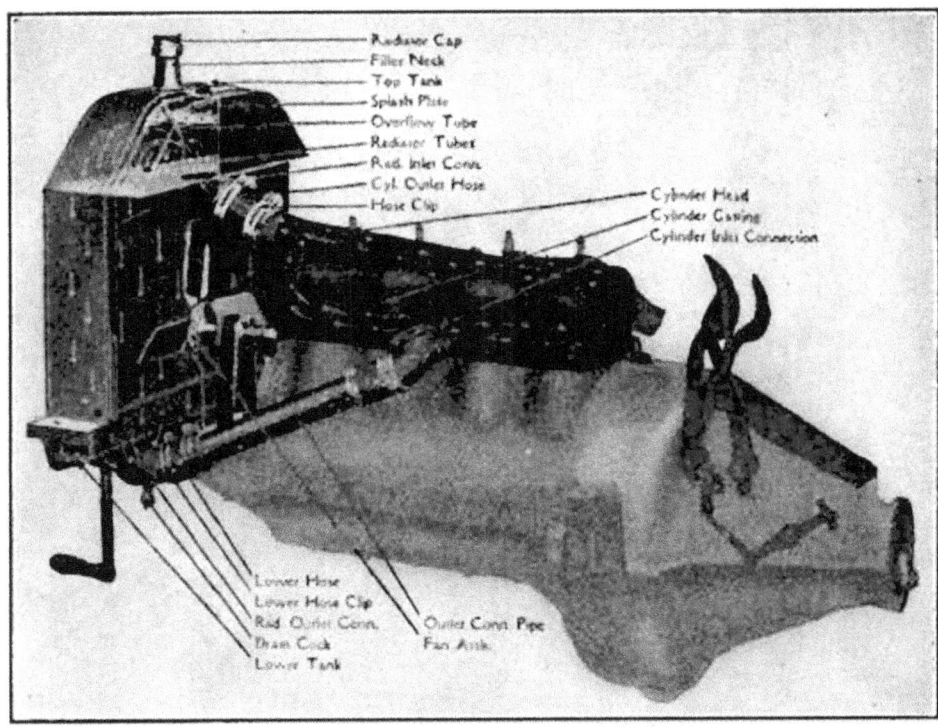

Fig. 127.—Ford cooling system.

through the lower water line into the cylinder water jacket, directly at the valve seats, where perfect cooling is most needed. Here it absorbs the heat and goes on to the upper water line and thence to the radiator. In the radiator D the water percolates slowly down through many fine tubes F and is cooled by the air rushing between the fins surrounding the tubes and thence returns to the pump. A fan G on the front of the motor, belted to the crank shaft, draws the air through the radiator and facilitates the cooling operation. Figure 128 also shows a standard design of tubular radiator. The pump, which is of the centrifugal type, requires no attention other than to see that it does not become choked by using dirty water. There is a packing nut on the shaft which should be repacked if the pump should ever leak around the shaft entrance.

This can very easily be done by turning off the packing nut, removing the old packing and rewinding the shaft with a few inches of well graphited packing and tightening up the packing nut. The packing should be wound on in the same direction as the nut is turned to tighten it.

The cooling system on the Cadillac Eight is of the forced circulation type. The radiator is of the tubular and plate type, with rotating fan mounted on the forward end of the generator driving shaft, the latter

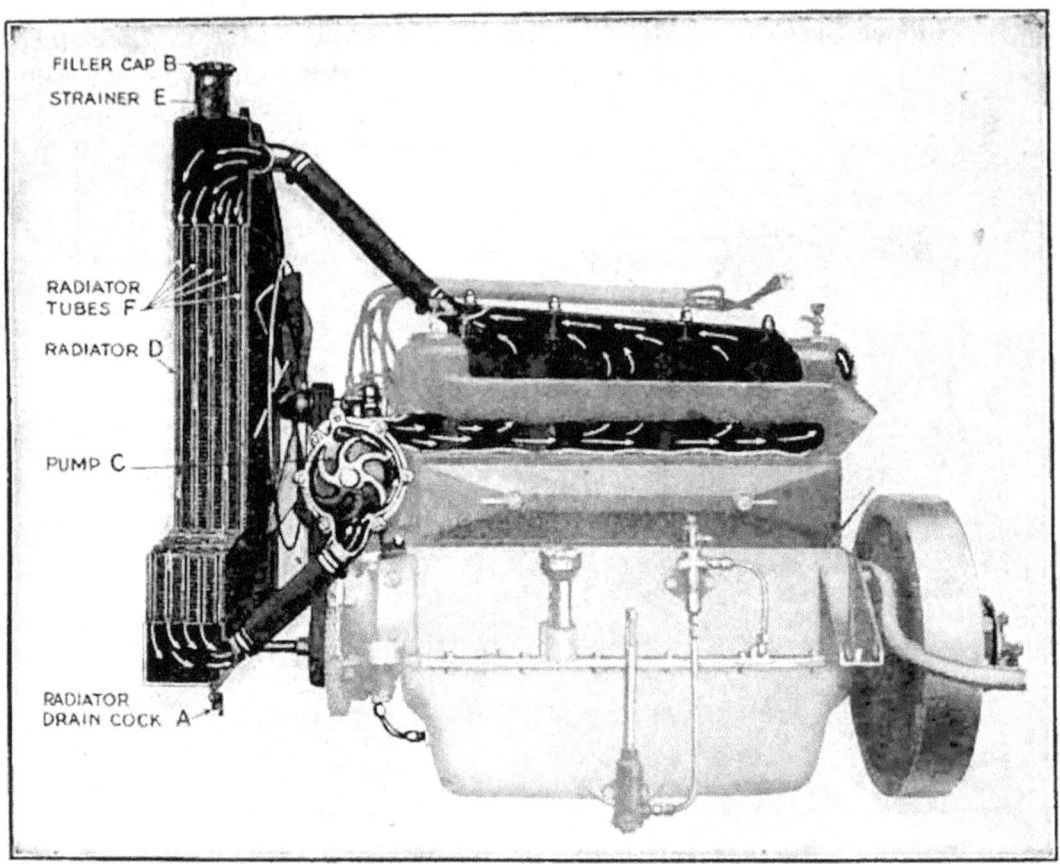

Fig. 128.—Studebaker cooling system.

being driven by silent chain from the cam shaft. Each set of cylinders is cooled separately. Due to the angle of jackets, the water does not lodge in the pockets. The natural tendency is for the water to flow upward and to rise to the hottest points.

There are two centrifugal water pumps, one on each side of the forward end of the engine. These are driven by a transverse shaft which is driven by spiral gears from the crank shaft. Within each pump housing is a thermostat shown in Fig. 129, which controls a valve that is between the radiator and the pump.

When the temperature of the cooling water drops below a predetermined temperature, the thermostats contract, thereby closing the valves. The water is then circulated only through the cylinder blocks and the carburetor jacket. It returns to the pumps through the water jacket on the intake manifold and carburetor. When the thermostats are closed, none of the water circulates through the radiator the evaporation of the gasoline in the carburetor and manifold providing sufficient cooling action. As the temperature of the water rises, the thermostats expand, thereby gradually opening the valves, permitting the water to circulate through the radiator.

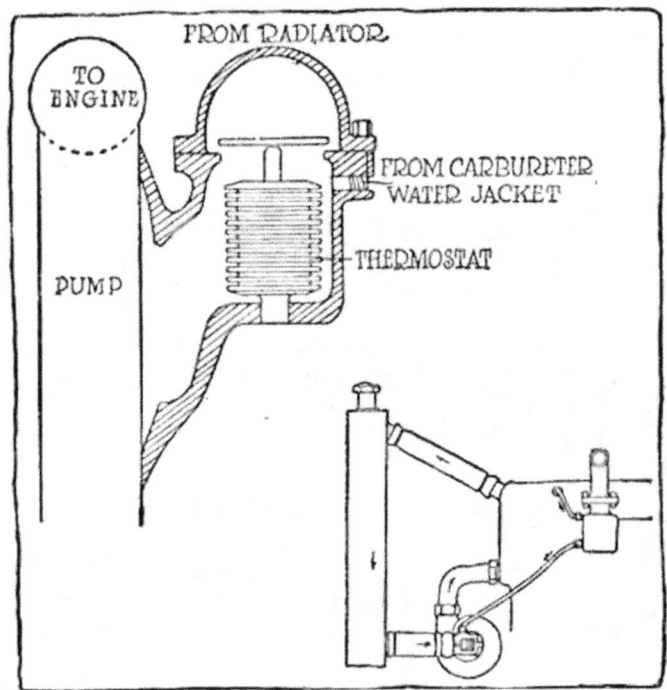

Fig. 129.—Cadillac thermostatic control of cooling water.

The advantage in this device is that, in starting with a cold engine, the engine is brought to a point of highest efficiency, in so far as heating is concerned, much more quickly than if it were necessary to heat the entire volume of water before reaching that efficiency. With the usual water circulating system, the highest efficiency of the engine is not reached in extreme cold weather. An engine uses its gasoline most economically when it is running rather warm, and with a radiator which is adequate to prevent overheating in hot weather, the cooling is too great for best economy in extreme cold weather.

The Cadillac thermostat is simply a small corrugated copper tube containing a liquid which expands or contracts in accordance with the temperature, thus slightly lengthening or contracting the tube, its total movement being $\frac{1}{4}$ in. This thermostat is in connection with a valve so that, when it expands, it raises the valve from its seat, this valve controlling the flow of water to the radiator from the pump. A by-pass connects with the water jacket of the carburetor, and when the engine is started, the water is naturally cold. Therefore the thermostat is contracted and its valve is seated. Thus the radiator water is shut off, the circulation being simply through the water jackets of the cylinders, through the by-pass to the carburetor jacket and thence back to the cylinders. There is thus only a small part of the water circulating, and when this heats up, the thermostat begins to expand and lifts its valve from its seat, letting the radiator supply flow into the system. This action continues back and forth so that the water temperature is nearly constant.

88. Air Cooling.—The Franklin engine, shown in Fig. 130, shows a good design of an air cooling system. The direct air cooling of the engine

Fig. 130.—Franklin air cooling system.

is accomplished as follows: The individual cylinders are provided with vertical fins projecting from their periphery. The fins on each cylinder are surrounded by sheet metal jackets which form passages for the air. The flywheel is provided with a number of curved blades so that it has a blower effect whenever the engine is running. This forms a partial vacuum and sucks air into the space underneath the hood through the grille in front. This air passes in uniform quantities down through the individual jackets on each cylinder into the compartment below the engine deck and hence out through the fan blades. The fan is incorporated in the flywheel and driven directly by the engine; so a steady stream of fresh air is being continually drawn down over the cylinders as long as the engine is running.

89. Cooling Solutions for Winter Use.—In climates where the temperature does not go below a dangerous freezing point, the cooling medium used is water; but in cold regions, where cars are run a good deal in the winter, it is necessary to get some kind of anti-freezing solution. The ideal requirements for an anti-freezing compound are as follows:

1. It should have no harmful effect on any part of the circuit with which it comes into contact.
2. It should be easily dissolved or combined with water.
3. It should be reasonably cheap.
4. It should not waste away by evaporation, that is, its boiling point should be as high as that of water.
5. It should not deposit any foreign matter in the jackets or pipes.

The principal materials used are: (1) oil; (2) glycerine; (3) calcium chloride; (4) alcohol; (5) mixture of alcohol and glycerine; (6) kerosene oil.

Oil has the advantage of having a very high boiling point so that it will not waste away, but it has the disadvantage that it does not make a good mixture with water, and will not absorb heat as rapidly as water. It also has a lower heat coefficient, that is, it takes less heat to raise the temperature of a certain amount of oil one degree, than it does the same amount of water. Oil cannot be used where there is any rubber in the circuit. It will attack rubber hose and gaskets very quickly and they will deteriorate rapidly.

The disadvantages of using glycerine are similar to those of the oil, chief of which is sure destruction to the rubber connection. It also is liable to contain free acids, and it is quite expensive.

Calcium chloride makes a very good solution with water, the freezing point depending upon the proportions used. The general solution is to use 5 lb. of the salt to 1 gal. of water. This solution will stand 39° below zero before freezing. It has the disadvantage of being very apt to cause electrolytic action where two metals are joined together. It is derived from hydrochloric acid, and is liable to contain free acids, which attack the metal very rapidly. Calcium chloride has the same appearance as chloride of lime, but has a somewhat different chemical composition. Pure calcium chloride is the only thing that can be used. The commercial chloride of lime sets up electrolytic action. The solution may be tested for acid by dipping a piece of blue litmus paper in it. If there is any acid present, the paper turns red. As the water is evaporated in the radiator there will be a crust formed on the inside of the jacket, and also in the pipes, which has a tendency to clog up and prevent circulation. The rate at which these deposits occur depends on the strength of the solution.

Denatured alcohol seems to be about the best substance to use as a non-freezing solution, as it has no destructive action whatever on either metal or rubber, makes no deposits and never causes electrolytic action. A solution of 50 per cent water and 50 per cent alcohol will stand about 32° below zero. The only disadvantage that it has is that it evaporates more readily than the water, so that when adding new solution, more alcohol than water must be added in order to keep the solution of the same strength. The combination of alcohol, glycerine and water seems to give very good results. In this combination, equal parts of alcohol and glycerine are used. The alcohol has a tendency to overcome the destructive action of the glycerine or the rubber connections, and the glycerine keeps the alcohol from evaporating too rapidly. The freezing point depends on the strength of the solution. A solution of 60 per cent water, and 20 per cent each of alcohol and glycerine freezes at 24° below zero. The proportions must be governed by the locality in which they are used.

There are also numerous anti-freezing compounds on the market. These are mostly put up from some of the materials mentioned here.

In the following tables are results showing the temperature at which some of the well known anti-freezing solutions will freeze, in various proportions of mixture with water and with one another. These are necessary, as different localities and different altitudes require different solutions and every person should be able to select his solution in the right proportion to avoid having any trouble in the coldest possible weather likely to be experienced in his home location.

FREEZING POINTS OF CALCIUM CHLORIDE SOLUTIONS

Per cent by volume of calcium chloride	Specific gravity of solution	Freezing point
10	1.085	22°F.
15	1.131	13°F.
20	1.119	0°F.
22	1.200	−9°F.
24	1.219	−18°F.
26	1.242	−28°F.
28	1.268	−42°F.

The specific gravity is given to be used as a check on the proportions.

FREEZING POINTS OF DENATURED ALCOHOL MIXED WITH WATER

Per cent by volume of alcohol	Specific gravity of solution	Freezing point
10	0.988	24°F.
20	0.975	14°F.
30	0.964	−1°F.
40	0.954	−20°F.
50	0.933	−32°F.
60	0.913	−45°F.
70	0.897	−57°F.

If wood alcohol be used instead of denatured alcohol, slightly lower temperatures can be reached with the same proportions of alcohol and water.

FREEZING POINTS OF ALCOHOL AND GLYCERINE MIXED WITH WATER

Alcohol and glycerine	Water	Freezing point
15 per cent	85 per cent	20°F.
25 per cent	75 per cent	8°F.
30 per cent	70 per cent	− 5°F.
35 per cent	65 per cent	− 18°F.
40 per cent	60 per cent	− 24°F.
45 per cent	55 per cent	− 30°F.
50 per cent	50 per cent	− 33°F.

CHAPTER VI

BATTERIES AND BATTERY IGNITION

All automobile engines in use at the present time have some form of electric ignition, in which a current of electricity is made to produce a spark inside of the cylinder. All ignition systems are made up of two essential parts: (1) the source of electric current supply; and (2) the apparatus for utilizing this current to produce a spark in the cylinder.

Before considering the features of either of these component parts it is necessary that an understanding be had of the fundamental electrical principles and definitions governing the construction and operation of electric ignition systems.

90. Fundamental Electrical Definitions.—An electric current flowing in a wire can be compared to water flowing in a pipe line. As the water pressure is measured in pounds per square inch, so the electrical *pressure* in a wire is measured by a unit called a "Volt." It is the practical unit by which electrical pressures are measured.

The "Ampere" is the practical unit by which the rate of current flow in a wire is measured. It corresponds to the number of cubic feet or gallons which flow through a water pipe per unit of time. For a large number of amperes, a large wire is necessary and for a smaller number of amperes, a smaller wire can be used. We can have a small wire carrying a current of high voltage, and a large wire carrying current of low voltage, just the same as a large or small pipe can carry water of either high or low pressure. The size of wire determines the quantity of current it can carry. A small wire can carry a small current but it requires a large wire to carry a large current.

The "Ohm" is the unit by which the resistance to the flow of electric current through a wire is measured. It corresponds to the friction opposing the flow of water through a pipe.

The Ampere-hour is the measure of quantity of current. One ampere-hour is the amount of current which would flow at the rate of 1 amp. in 1 hour. It is by this unit that the capacity of storage batteries is measured. A 60 ampere-hour battery will give current at the rate of 60 amp. for 1 hour, or at the rate of 30 amp. for 2 hours, or at the rate of 1 amp. for 60 hours, etc.

91. Direct and Alternating Current.—Electric current can be of two kinds: direct or alternating. Direct current always flows in one direction in the wire, and is the kind of current which is given out by every type of battery. Alternating current, however, first flows in one direction and then in the other, the reversals taking place many times per second. It is the kind of current given out by most of the modern magnetos.

92. Dry Batteries.—The first necessary part of an electric ignition system is the source of current. For this purpose we can have either batteries, dynamos, or magnetos. In this chapter only batteries and battery ignition systems will be discussed. Magnetos will be treated in the chapter on Magnetos.

The dry battery is a common source of battery current for ignition purposes. It is comparatively cheap, exceptionally reliable, and can be easily replaced when worn out. Due to improvements in the battery ignition systems its use for motor car ignition is growing, after having given way for a time almost entirely to magneto ignition. Figure 131 is a section of a commercial dry cell. It consists of a cylindrical zinc shell around the inside of which has been placed a piece of absorbent paper saturated with a paste made of zinc oxide, zinc chloride, ammonium chloride, plaster of Paris, and water. The zinc can forms the negative terminal of the battery, and the carbon element down through the center of the cell forms the positive terminal. The space between the absorbent paper and the carbon is filled with powdered carbon and manganese oxide which acts as a depolarizing agent. The voltage of a dry cell is about 1.5 volts. The maximum possible amperage or current of a new cell ranges from 20 to 35 amp., depending upon the size of the cell. The dry battery always gives out direct current. The capacity and life of a dry cell depends on the way it is used, being greater when it is used intermittently.

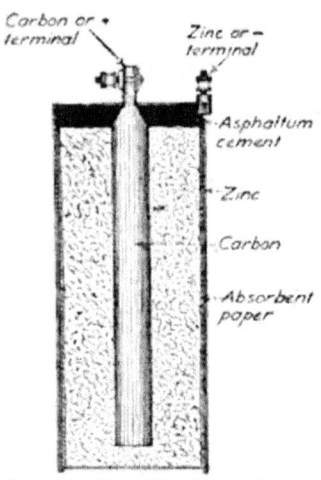

Fig. 131.—Section of dry cell.

93. Storage Batteries.—Although the storage battery is to be considered in Chap. VIII on Starting and Lighting Systems, a brief description will be given here in order to bring out clearly its functions in battery ignition systems. A storage cell, Fig. 132, consists of two sets of metallic plates placed in a vessel containing a solution of sulphuric acid and water. In the positive group the plates are lead grids, the openings being packed with lead peroxide, characterized by its chocolate brown color. The plates of the negative group consist of finely divided sponge lead. These sets of plates are placed in the cell so that the positive and negative plates alternate and are separated by perforated sheets of hard rubber or specially treated wood. By passing direct current into the top of one of the plates, through the acid and water, and out the other plate, the plates are changed chemically. When the battery is used, the chemical change is reversed and the plates tend to return to their original state, giving off current as they do so. The single storage cell of one positive and one negative set of plates gives, when fully charged, a pressure of about 2 volts and a current depending upon the size and number of the plates. For ignition purposes the plates are connected so that the whole battery gives a voltage of from 6 to 8 volts and a capacity of from 60 to 80 ampere-hours.

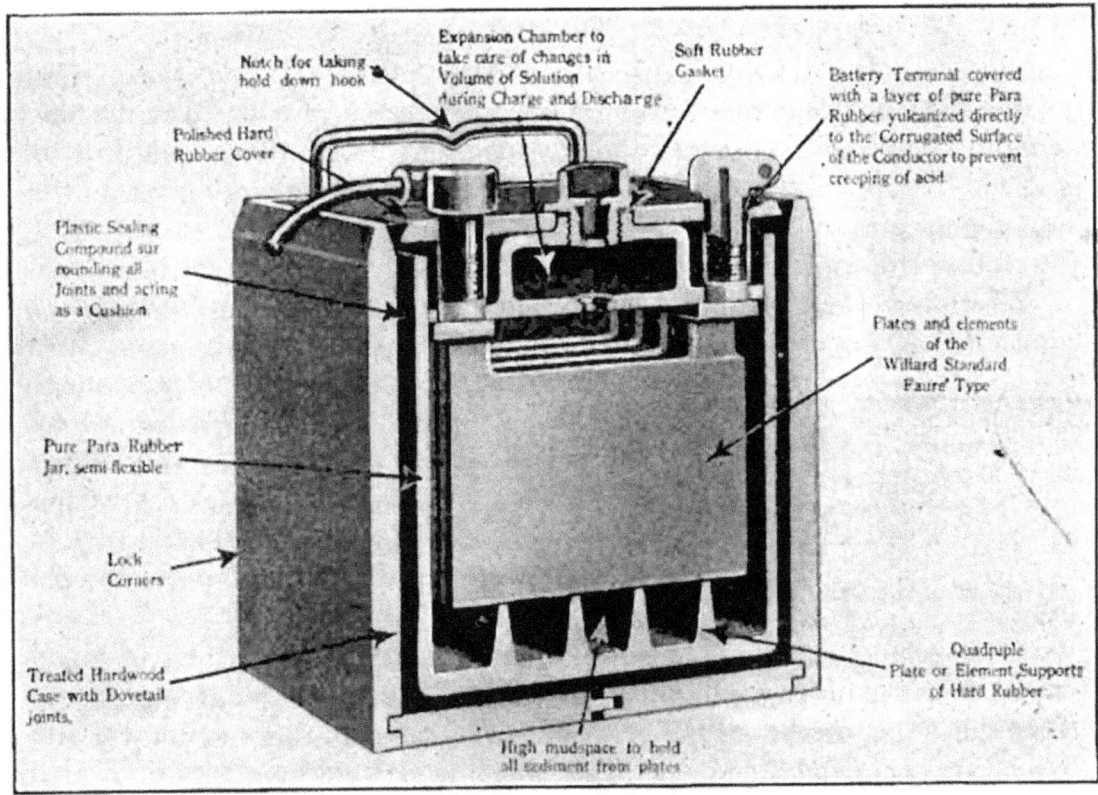

Fig. 132.—Section of Willard storage cell.

94. Series and Parallel Connections.—The voltage of either a dry or storage cell is not high enough for automobile engine ignition purposes, and methods of connecting several batteries must be resorted to in order to raise the voltage and amperage. A voltage of from 6 to 8 is necessary for an ignition system using an induction coil. This can be obtained by the connection shown in Fig. 133, in which the carbon of one cell is connected to the zinc of the next. This is known as the "series" connection. By so connecting the cells, the resultant voltage is equal to the combined voltage of all, or the number of cells multiplied by the voltage of one cell, which is 1.5. The current output is equal to the current of one cell of the given size, or about 20 amp. If all the carbons are connected and all the zincs fastened together, as shown in Fig. 134, the connection is known as "parallel." The resultant voltage equals the voltage of one cell and the current output equals the current output of one cell multiplied by the number of cells. Therefore, to increase voltage connect the cells in series, and to increase current output connect them in parallel.

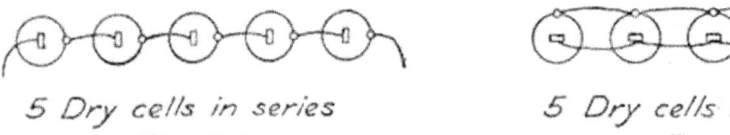

5 Dry cells in series
Fig. 133.

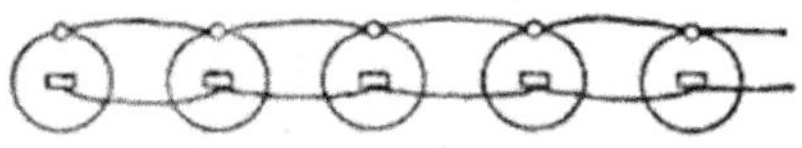

5 Dry cells in parallel
Fig. 134.

95. Battery Connections for Ignition Purposes.—Where the current demand is small or not continuous, a single series of cells (usually five) is used. This arrangement is suitable for single cylinder engines, or for starting engines of two or more cylinders, where a magneto is used after the engine is in operation.

When the amount of current required is great, the multiple series connection is used. It is suitable for engines of two or more cylinders and continuous service. This arrangement consists of parallel groups of as many cells in a series as may be required for the service. Figure 135 shows an arrangement with three parallel sets, each of five cells connected in series. This arrangement provides for an amperage of about 60 at about $7\frac{1}{2}$ volts.

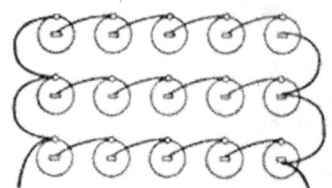

15 cells in multiple series arrangement
Fig. 135.

Two series of cells in multiple series connection will have about three times the life of a single series on the same current, on account of the reduced rate of discharge. Three series connected in this manner will give about six times the life on the same current, as would one series.

Another advantage of this method of connection is that a dead cell will not weaken the current from the group enough to interfere with the engine operation. For ordinary service, three groups of five cells each are frequently used, while for heavy, constant service five groups of five cells each, giving a voltage of about 7.5 and a current of about 100 amp. is recommended.

96. Simple Battery Ignition System.—The jump-spark or high-tension system of ignition is so named because a high tension current is caused to jump across the gap between the terminals of the spark plug in the cylinder. Figure 136 shows an elementary battery jump-spark ignition system for a one-cylinder engine. Four dry cells are shown connected in series giving a voltage of about 6 and a current of about 20 amp. One terminal of the battery set is connected to the left terminal of the induction or spark coil and the other terminal to the engine "timer." The timer, or commutator, is nothing more or less than a mechanically operated "switch," placed between the batteries and the right terminal of the coil. The current from the batteries goes to the left terminal of the coil which is connected to a standard holding an adjustable contact screw. This screw is in contact with the vibrator. Passing from the screw into the vibrator, the current goes through a comparatively large wire wound around the central core. This wire goes to the right terminal of the coil, which is connected back to the timer. This circuit forms what is known as the "primary" of the system. When the timer completes the circuit,

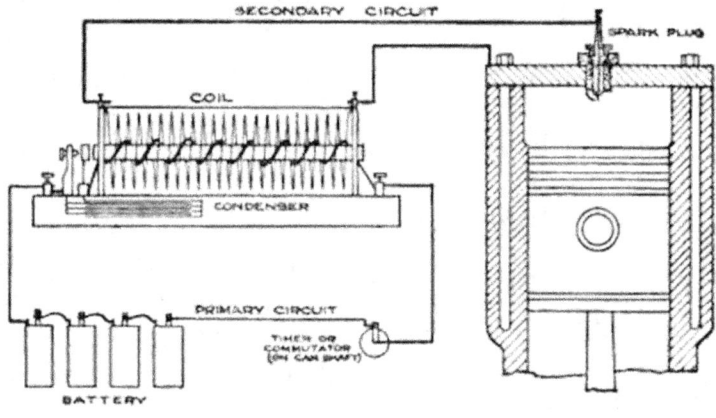

JUMP SPARK SYSTEM OF IGNITION
Fig. 136.

current flows through the primary winding. The current flowing around the iron core makes a magnet of it. This fact causes the vibrator to be pulled away from the adjusting screw, and this breaks the circuit. Consequently, current ceases to flow, the core loses its magnetism, the vibrator flies back to make contact with the screw again, and this permits the primary current to flow, causing a repetition of events. The result is a constant dying down and building up of the current in the primary winding around the core. This results in a dying down and building up of the magnetism in the core. It will be noticed that there is another coil of finer wire wound around the primary coil on the iron core. This is called the "secondary" of the coil. The ends of this secondary winding are fastened to the two secondary terminals on the top of the coil. One terminal of the coil is connected to the spark plug in the cylinder and the other is connected onto the engine frame, or "grounded."

Each time the current in the primary circuit is *broken*, there is another current of very high voltage induced in the secondary winding. This current is of sufficiently high electrical pressure to jump the spark plug gap under the usual compression pressure. This voltage varies from

10,000 to 20,000 volts. The relation between the voltage on the primary circuit and that on the secondary depends upon the relative number of turns of wire on the primary and secondary windings, upon the speed of the vibrator and the current in the primary winding.

In the bottom of the coil is placed the *condenser*, consisting of alternate tinfoil and oiled paper sheets. Every alternate tinfoil sheet is connected to the bottom of the standard; the ends of the others are connected to the vibrator. The current tends to continue flowing after the circuit is broken and, if it were not for this condenser, there would be a fat spark across the vibrator points every time the circuit was broken. The condenser prevents this arcing across the vibrator points, when they break, by absorbing this flow of current and storing it until the circuit is again closed. In addition, it aids in the induction of the high tension current in the secondary winding of the coil by permitting the quick break of the primary current.

The following are the names and functions of the various parts of a battery ignition system:

Primary Circuit.—That part of the system carrying the battery current at low voltage—a few turns of coarse wire on the coil.

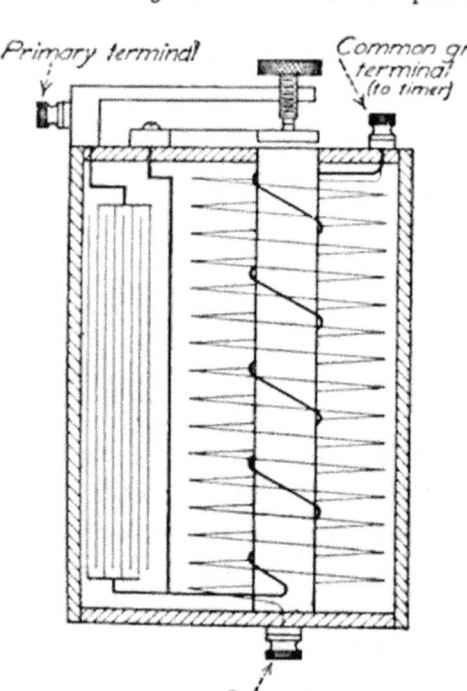

Fig. 137.—Three terminal vibrating induction coil.

Secondary Circuit.—That part of the system carrying the high tension current to the spark plugs—a great many turns of very fine wire on the coil.

Timer.—A mechanically operated switch placed in the primary circuit. Its function is to complete the primary circuit and cause the vibrator to act, thus causing a high tension current to flow to the spark plugs at the proper time.

Vibrator.—A spring placed in the primary circuit to make and break the current, causing a high tension current in the secondary.

Condenser.—An electrical appliance placed in the primary circuit to prevent sparking at the vibrator points.

97. The Three Terminal Coil.—Most of the coils used on automobile ignition systems have only three terminals instead of four. One end of the primary winding is joined to one end of the secondary and the junction to one of the terminal binding posts. The other end of the primary goes to a primary binding post and the other end of the secondary to the secondary binding post of the coil as shown in Fig. 137. An

Fig. 138.—Pfanstiehl three terminal coil.

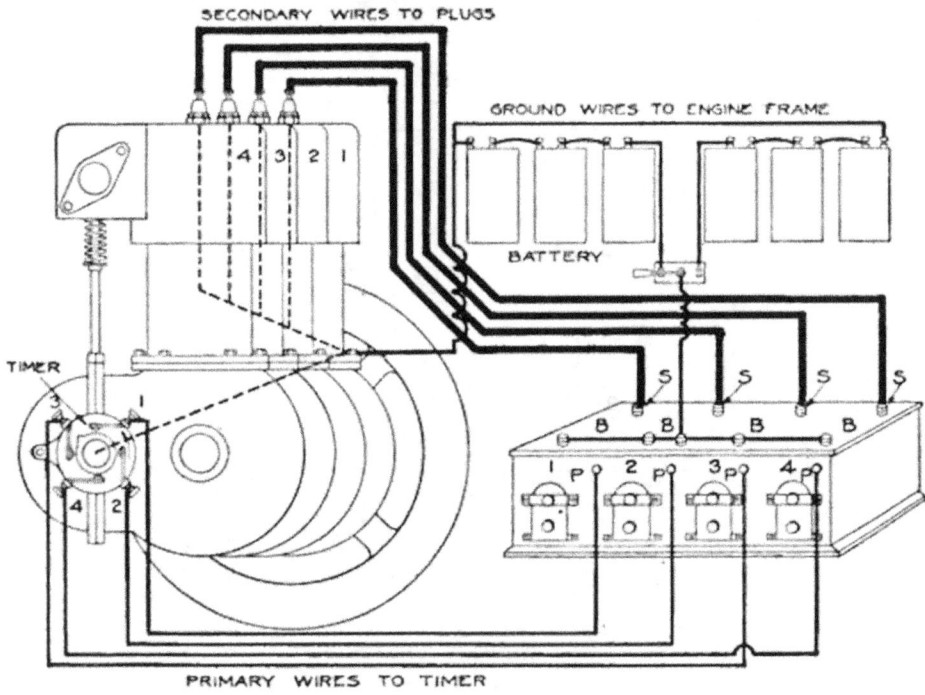

Fig. 139.—Wiring diagram for four-cylinder engine.

external view of a three-terminal coil for a single-cylinder engine is shown in Fig. 138. In Fig. 139 a four-unit coil with the wiring for a four-cylinder engine is shown. The three terminals are lettered: S, the secondary terminal leading to the plug; P, the primary terminal to the timer; and B, the terminal connected to the batteries. The secondary circuit is from the secondary terminal to the plug, across the gap into the engine frame, back through the timer to the coil. The primary circuit is from the batteries, one side of which is grounded, through the coil, to the timer, where the circuit is grounded and the current returns to the batteries through the metal of the engine.

FIG. 140.—Pfanstiehl four-cylinder coil set.

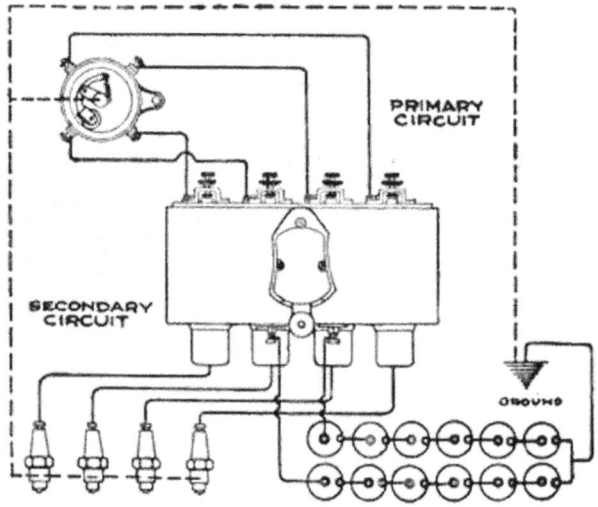

WIRING DIAGRAM FOR 4-CYLINDER ENGINE USING DRY CELLS

FIG. 141.

Where a multiple cylinder engine is used, it is customary to use a coil for each cylinder. The coils are usually enclosed in an upright box as shown in Fig. 140, which is a coil set for a four-cylinder engine.

In Fig. 141 is shown the arrangement of the ignition system for a four-cylinder engine using dry batteries as the source of current. There are two sets of batteries, one service set and a reserve set. The six cells are connected in series, giving a voltage of about 9. The four coils are placed in one box, with two small terminals at the bottom. Either of these terminals is a primary terminal for any one of the four coils and is connected to the two sets of batteries. The switch on the front of the box determines which set of batteries will be used. The other primary terminals at the top of the coils are connected to the four binding posts of the timer. These terminals are also secondary terminals. The large connections at the bottom of the coil box are secondary wires leading to the spark plugs. When the timer, which runs at one-half engine speed or at cam shaft speed, grounds the primary circuit by the roller making contact with the insulated terminal, a spark occurs in one of the cylinders, depending upon the position of the roller. Any of the four coils may be removed from the box for adjustment or repair.

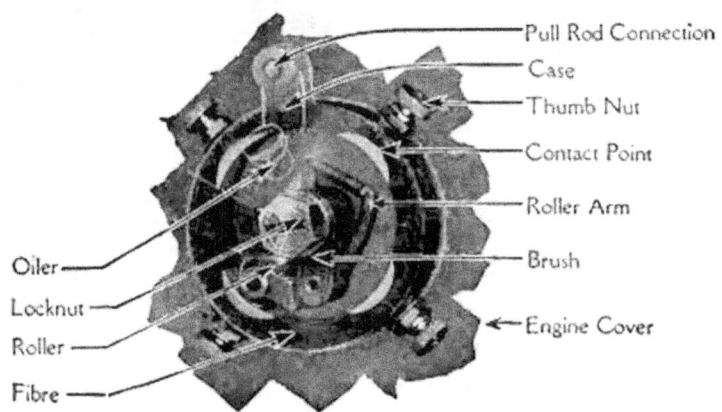

Fig. 142.—The Ford timer.

98. Timers.—Figure 142 shows the timer used on the Ford engine. The inside or rotating part is fastened to and rotates with the cam shaft. When the roller comes into contact with one of the terminals on the housing, the circuit for that coil is closed and a current is caused to flow in the primary circuit, causing a spark in the secondary circuit. The housing does not turn with the cam shaft, but can be shifted back and forth, either advancing or retarding the spark. The timer is always placed in the primary circuit.

The timers for six- and eight-cylinder engines are similar to the above, but have six or eight insulated terminals on the housing instead of four.

99. Spark Plugs.—The spark plug consists of two terminals fastened together, but insulated from each other, and the whole screwed into the cylinder. The center terminal is insulated from the rest of the plug and the other terminal. The insulation between the center electrode and the body of a plug is usually either of porcelain or of mica. The outside terminal is in contact with the engine cylinder and is consequently grounded. The only way the current can get from one terminal to another is across the air gap between them. The gap between points of the battery spark plugs should be about $\frac{1}{32}$ in., or the thickness

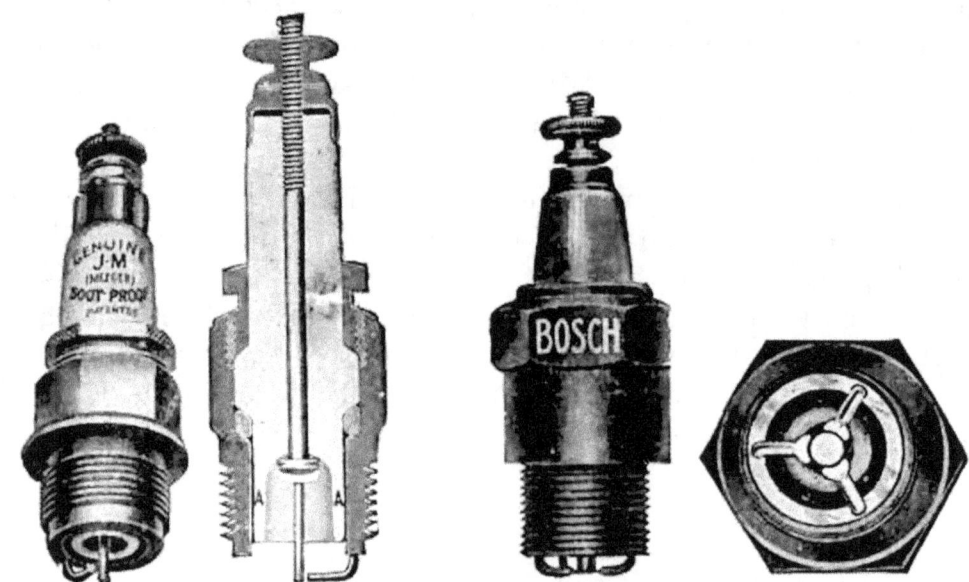

Fig. 143.—J. M. soot-proof spark plug.   Fig. 144.—Bosch spark plug.

of a smooth dime. Figure 143 shows the exterior and interior arrangement of the J. M. soot-proof plug. In Fig. 144 is shown the side and bottom views of the Bosch plug with three grounded electrodes.

Fig. 145.—Pfanstiehl master vibrator.

100. Master Vibrators.—In order to avoid the four vibrator adjustments on the four-coil systems, and the possibility of getting sparks of different intensity in the different cylinders, a master vibrator is sometimes used. A master vibrator is an additional coil with only a primary winding, one vibrator, and a condenser. It is placed between the batteries or source of current and the primary windings of the coils. The vibrators of the coils are then screwed down tight or short-circuited by a copper wire as shown in Fig. 146. The master vibrator serves for all four coils and, when once adjusted, the sparks in all the cylinders will be of the same intensity. There is only one vibrator to be adjusted and to get out of order instead of four. The principle of the master coil is that the winding of the coil and the vibrator are connected successively in series with the primary windings of each individual coil.

This produces the make and break in the primary winding of the coil. Figure 145 illustrates the outside view of the Pfanstiehl master vibrator. Figure 146 shows the application of the K-W master vibrator with both battery and magneto sources of current.

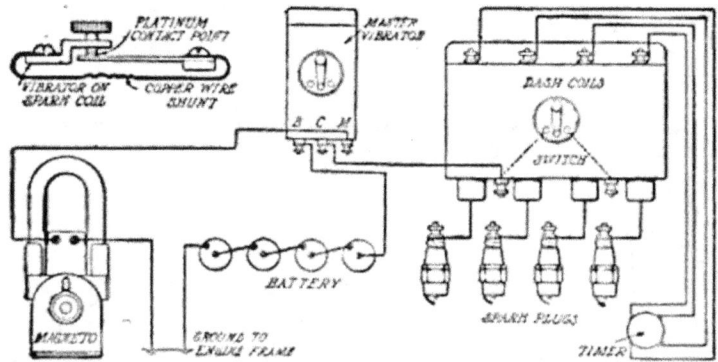

Fig. 146.—Connections for K-W master vibrator.

101. The High Tension Distributor System.—A typical high tension distributor system is shown in Fig. 147. This system enables a single coil to be used to serve a number of cylinders. The particular feature of this system is the combined low tension timer, or interrupter, and the high tension or secondary distributor, acting with a single non-

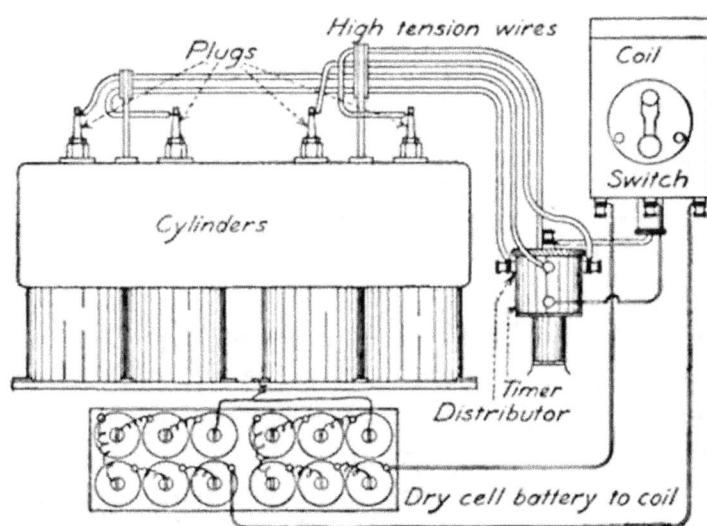

Fig. 147.—High tension distributor system.

vibrating coil. In this particular illustration, two sets of dry cells are provided, one set being in reserve. The distributor and timer are usually mounted in a vertical position in a single unit and are driven at cam shaft speed by a vertical shaft. The coil, as mentioned before, is non-vibrating. The mechanical contact maker, or interrupter form of timer located under the high tension distributor, serves in place of the usual vibrator on the coil.

The primary current flows out of the batteries into the bottom primary terminal of the coil, out of the center primary terminal and over to the primary binding post on the timer. The revolving contact maker completes the circuit by grounding the current through the timer shaft. This contact maker or timer is constructed so as to give a very quick break to the primary circuit so that there will be a high pressure current induced in the secondary winding of the coil. This flows out of the

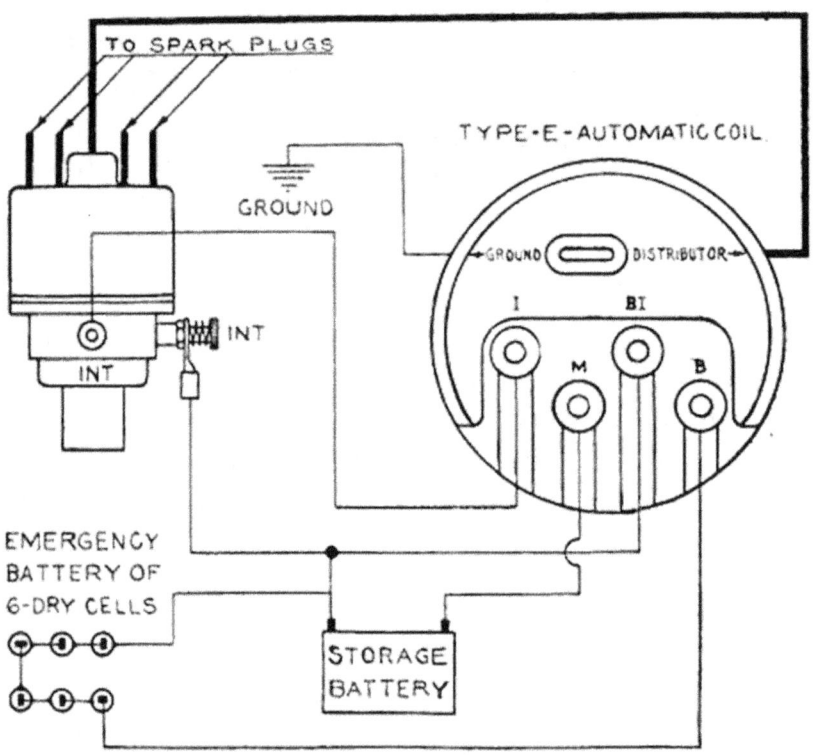

Fig. 148.—Connecticut type *E* ignition system.

secondary terminal of the coil to the main terminal post of the distributor, where it is sent to one of the four spark plugs, depending on the position of the distributor arm. The action of a distributor is much like that of an ordinary timer used with vibrating coils, though its construction to handle secondary high tension current is necessarily much different. Instead of producing a series of sparks in the cylinder, as is done with the vibrating coil, the mechanical interrupter produces only one fat spark in each cylinder.

This arrangement is not so complicated as the multiple coil system. There is only one adjustment, that at the contact maker, and this insures sparks of the same intensity in each of the cylinders. The drain on the batteries is also less, as only one spark is produced in each cylinder, in contrast to the series of sparks produced by a vibrating coil.

102. The Connecticut Automatic Ignition System.—This system operates on the high tension distributor principle, using but one coil for all cylinders. It employs a mechanical interrupter for the primary current. Although dry batteries can be used in cases of emergency, the system is primarily intended for the use of storage batteries as the source of current. Its ideal use is in conjunction with a generator supplying current to a storage battery for lighting and starting. Figures 148 and 149 are wiring diagrams showing the connections for the Connecticut

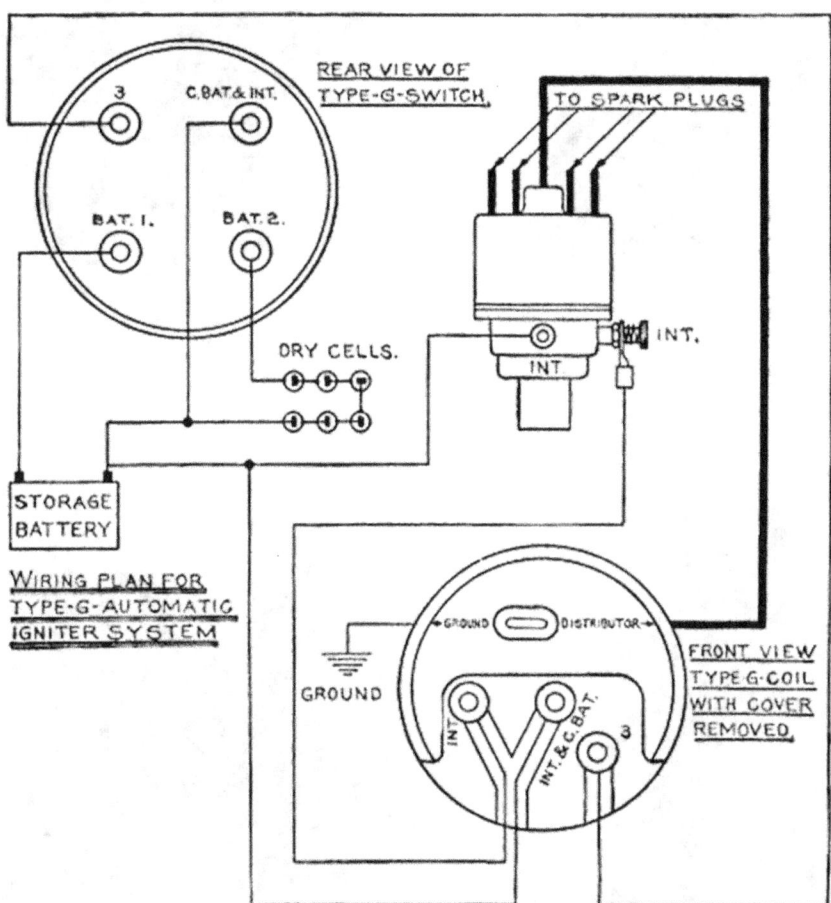

Fig. 149.—Connecticut type G ignition system.

types E and G systems. The essential difference between these systems is in the switch and coil connections. In type E the coil is integral with the switch and is designed to be placed under the hood, thus assisting in preserving a clean dash. Type G has a separate switch and coil, which permits its application where the space is limited, as for instance when the gasoline tank is carried in the cowl dash. The switch is mounted on the dash and the coil any place on the engine near the igniter, thus bringing the condenser close to the breaker points and eliminating the necessity of extending the high tension wires to the dash.

The combined interrupter and high tension distributor is clearly shown in Figs. 150, 151, and 152. The interrupter, Fig. 150, first closes the circuit and permits battery current to flow through the primary circuit. When one of the lobes on the cam strikes the roller, the circuit

Fig. 150.—Connecticut interrupter.

Fig. 151.—Connecticut igniter with distributor cap removed.

Fig. 152.—Connecticut igniter assembled.

Fig. 153.—Connecticut type *E* coil and switch with cover removed to show terminal connections.

is opened and a high voltage is thus produced in the secondary winding of the coil. The high tension current is distributed to the plugs by the distributor of the instrument. The distributor and interrupter are mounted in a single unit as shown in Fig. 152, the whole device being called the igniter.

The igniter is mounted on a vertical shaft running at one-half engine speed and thus can be mounted the same as the ordinary timer for vibrating coils. Figure 153 shows the arrangement of the coil terminals. It will be noted that a spark gap is provided to protect the secondary winding from the destructive action of the high voltage in case a plug terminal becomes disconnected so that the high tension current can not take its regular path. The safety gap is placed in a glass tube inaccessible to vapor or fumes. It is conveniently arranged for observation in cases of missing cylinders.

The spark advance and retard in the Connecticut system are effected by swinging the entire igniter housing either forward or back.

103. The Atwater Kent System.—The Atwater Kent system is also of the high tension distributor type and has as an optional feature the

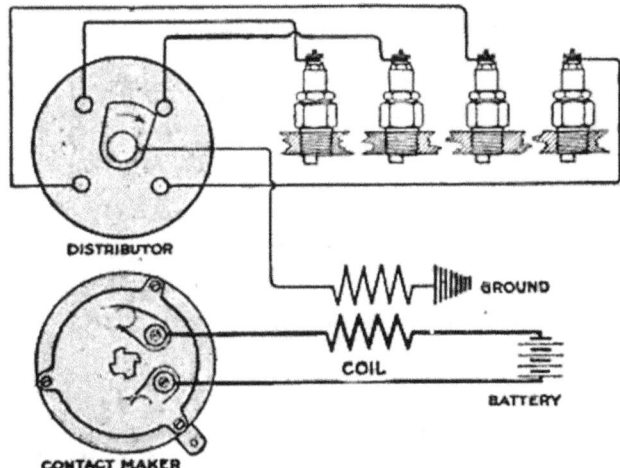
Fig. 154.—Diagram showing principle of Atwater Kent system.

Fig. 155.—Exterior of unisparker.

automatic spark advance, which automatically regulates the position of the spark according to the speed of the engine. This system is designed to operate in a satisfactory manner with dry cells as the source of current.

The Atwater Kent system consists of two main parts:

(1) The unisparker, which is the contact maker and the distributor combined in one small case mounted on the timer shaft of the engine.

(2) The coil, which consists of a simple primary and secondary winding with condenser. The coil has no vibrators or other moving parts, this function being served by the contact maker. The principle of the Atwater Kent system is clearly shown in Fig. 154. The battery current is closed and broken by the mechanical contact maker. The secondary current from the coil goes to the distributor, where it is

directed to the proper plug. The distributor and contact maker are built together and are called the unisparker.

The unisparker, Type K-2, is illustrated in Fig. 155. It is connected to the ordinary timer shaft of the engine, the dome-shaped cover containing the primary contact maker and the secondary distributor as well as the spark advancer. By releasing the two spring clips, the

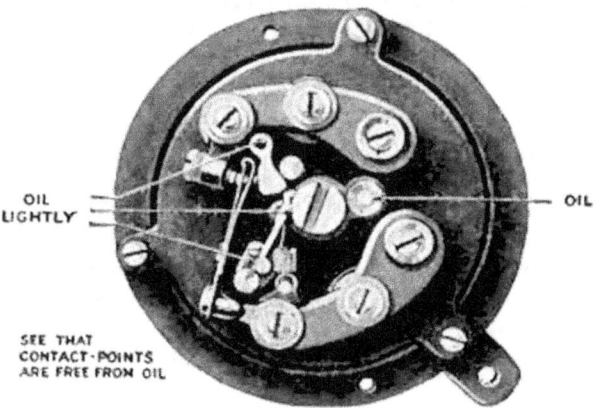

Fig. 156.—Atwater Kent contact maker.

rubber dome is lifted and the contact maker exposed. The contact maker of the unisparker is shown in Fig. 156. As will be seen from investigation, only one spark is produced per explosion stroke, as the circuit is made and broken but once. An important feature of this contact maker is that the length of contact is absolutely independent of the engine speed, and as strong a spark is produced when the engine is cranked by hand as when the latter runs at normal or even at racing speed. The length of

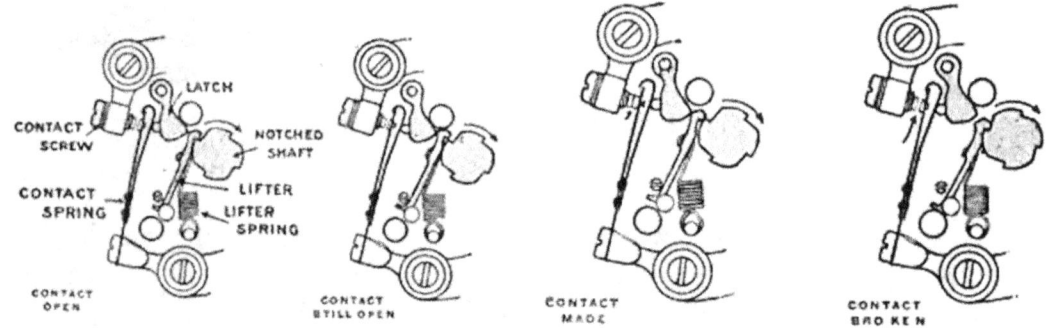

Fig. 157.—Operation of contact maker.

contact is constant and no greater at any speed than is necessary to insure the magnetic field of the coil being built up to its full strength.

The action of the contact maker is shown in Fig. 157. The hardened steel rotating shaft in the center, the lifter, the latch, and the contact spring are the principal moving parts. The contact is made and broken by the action of the lifter spring in drawing the lifter back, after it has become unhooked from the notched shaft. This spring action makes the speed of the break independent of the speed of the engine. It also makes the time of contact uniform, and this is adjusted so as to use the least possible current from the batteries.

Directly above the contact maker is located the high tension distributor. This consists of a revolving hard rubber block driven by means of a key from the end of the operating shaft and carrying a contact segment on its circumference. Two, four or six contact pins, depending on the number of cylinders, are secured into the hard rubber cover plate of the device, which, as already stated, forms the body of the distributor. Proper cable connections are formed on the terminals of

FIG. 158.—Atwater Kent kick switch coil.

the cover plate and, from these, connection is made to the individual spark plugs.

The coil used in connection with the Atwater Kent system consists of simple primary and secondary windings of generous proportions, which,

Motor stopped or running slowly.   Motor at high speed.
FIG. 159.—Atwater Kent automatic spark advance mechanism.

together with a condenser, are sealed into a container. There are no moving parts or adjustments.

One of three types of coils is usually furnished with the Atwater Kent system: a simple plate switch coil, a kick switch coil, shown in Fig. 158, or an underhood coil with separate switch. Both plate and kick switches are provided with a push button for producing starting sparks without cranking.

Automatic Spark Advance.—Figure 159 shows the centrifugal governor which advances the spark as the speed increases. The rotating shaft is divided, and as the governor weights expand they rotate the upper part of the shaft forward in its own direction of rotation, thus making and breaking contact earlier than at slow speed.

In Fig. 160 the wiring diagram of the Atwater Kent installation is shown. Among the particular features of this system are: time of closed primary circuit is independent of engine speed; speed of break is independent of engine speed; circuit cannot be closed when engine is stopped; battery consumption is reduced to a minimum; the spark is uniform in all cylinders and is independent of engine speed.

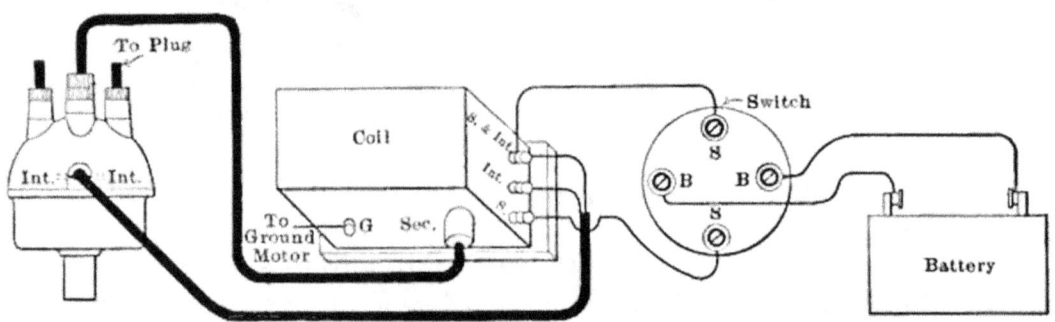

FIG. 160.—Atwater Kent wiring diagram.

104. The Westinghouse Ignition System.—There are several ignition systems made, particularly for cars equipped for electric starting and lighting, in which the source of current is a storage battery kept charged at all times by the starting and lighting generator. In some, the generator simply keeps the battery charged and the ignition system is entirely separate but draws its current from the battery. In others, the generator carries the interrupter and the high tension distributor for the purpose of timing and distributing the current.

The Westinghouse system of ignition is mounted as a unit with the electric generator which supplies electric current to the storage battery for lighting or starting or both. When the engine is not running or is operating at very low speed, the ignition current is supplied entirely by the battery. After the engine reaches a certain speed, the current may be supplied in whole or in part by the generator.

The ignition outfit consists, in addition to the generator and storage battery, of an ignition switch and coil on the dash, and an interrupter and distributor which are made a part of the generator. The ignition coil transforms the voltage of the battery up to the high tension required for the spark plugs. The interrupter closes and then opens the ignition circuit at each half revolution of the generator shaft, and the distributor directs the high tension current to each of the spark plugs in succession. Figure 161 shows the exterior of the generator with the distributor and interrupter on the right hand end.

Fig. 161.—Westinghouse ignition and lighting generator.

The view of the generator disassembled, Fig. 162, shows the principal parts. This system has an automatic spark advance operated by centrifugal weights inside the interrupter.

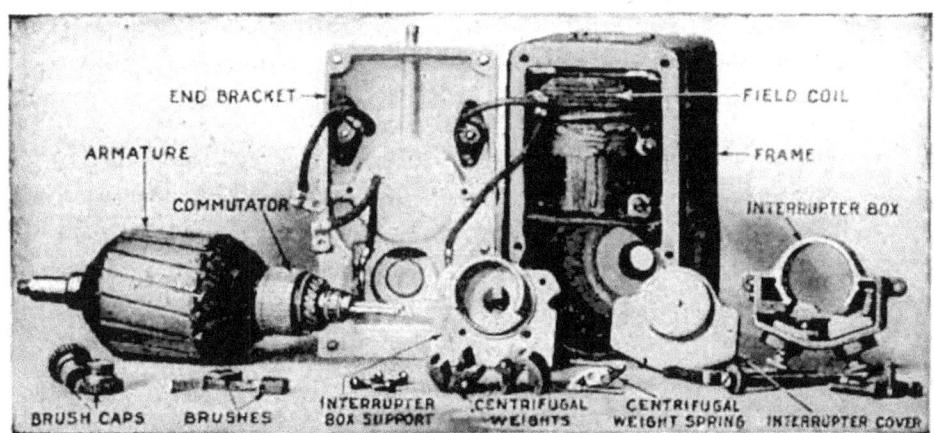

Fig. 162.—Parts of Westinghouse ignition and lighting generator.

Figure 163 illustrates the interrupter with the centrifugal weights and springs in the position they occupy when the engine is at rest. Figure 164 shows the position that the weights occupy when the engine is running at high speed.

The operation of the ignition system, including the interrupter and distributor, ignition coil and switch, begins with the "making" of the primary circuit of the coil when the centrifugal weights push down the

Fig. 163.—Westinghouse generator with distributor and interrupter cover removed.

fiber bumper, forcing the interrupter contacts to close. Then the weight moves off the fiber bumper, allowing the contacts to suddenly separate or open. This break of the primary circuit induces a high voltage in the secondary of the ignition coil. This is led to the distributor, which directs it to the proper spark plug, causing a spark at the spark plug gap. As the speed of the engine increases, the weights are thrown out from the center and automatically advance the time of closing or opening the interrupter contacts, and hence advance the spark. At the same time, due to their shape, they keep the contacts closed during a longer period of the revolution when running at high speed; this makes the

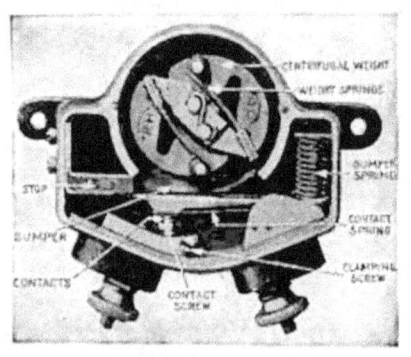

Fig. 164.—Westinghouse interrupter, showing position of weights at high speed.

length of time of contact practically the same at all speeds and prevents the spark voltage from falling off at high speeds.

In generators not provided with automatic spark advance the centrifugal weights are omitted and a solid cam substituted. The interrupter contacts are changed so as to make the breaking of the contact occur when the lever is pushed down by the cam instead of when being returned by the spring.

105. The Delco System of Ignition.—All the Delco systems are not identical, there being slight changes to adapt them to the different cars. For example, the ignition coil on some cars is mounted on the dash, or on top of the starting motor-generator instead of on the side, as shown in Fig. 165.

All current for lights, horn, and ignition is supplied first to the combination switch, and after passing through the protective circuit breaker

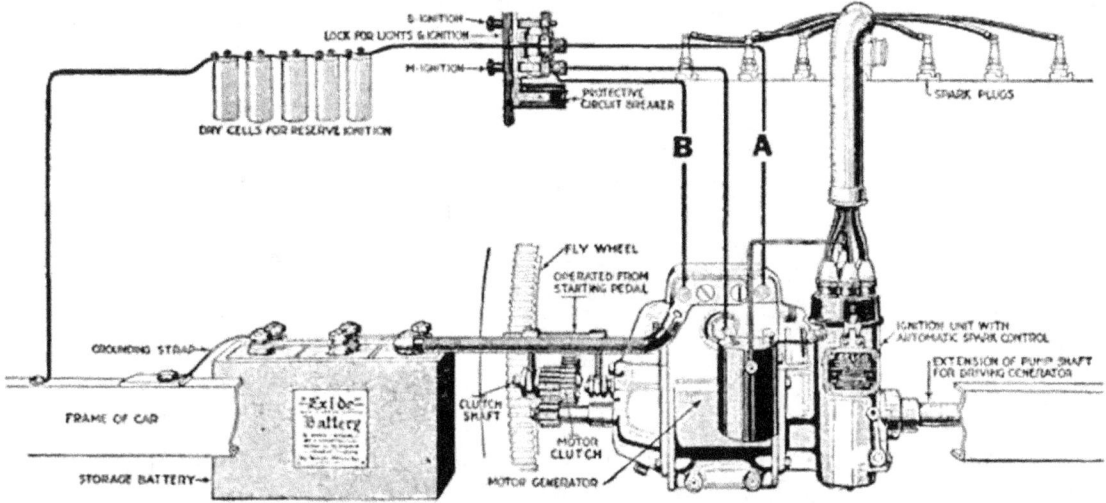

Fig. 165.—Delco ignition system.

on the dash is distributed to these different units. When the generator is supplying the current, it comes from the forward terminal on the side of the generator through the wire A to the switch. The storage battery current is connected through the wire B. If the button B is pulled out, the current from the dry cells is used for ignition. If button M is pulled, the current will be taken either from the generator or storage battery, depending on whether or not the generator is in operation. Thus, either the M or B button may be used for starting.

The excess current from the generator flows through the wire B to the storage battery. An ammeter inserted in the line A would indicate the amount of current coming from the storage battery to the generator when the engine is not running, or it would indicate the current being generated when the engine is running.

Distributor and Timer.—The distributor and timer is carried on the front of the motor-generator, and is driven through a set of spiral gears attached to the armature shaft. The distributor consists of a cap or head of insulating material, carrying one high tension contact in the center, with similar contacts spaced equidistant about the center, and a rotor which maintains constant communication with the central contact. The rotor carries a contact button which serves to close the secondary circuit to the spark plug in the proper cylinder.

Beneath the distributor head and its rotor is the timer, a diagram of which is shown in Fig. 166. This is provided with a screw A in the center of the shaft, the loosening of which allows the cam to be turned in either direction to secure the proper timing, turning in a clockwise direction to advance and counter-clockwise to retard. The spark occurs at the instant the timer contacts are opened.

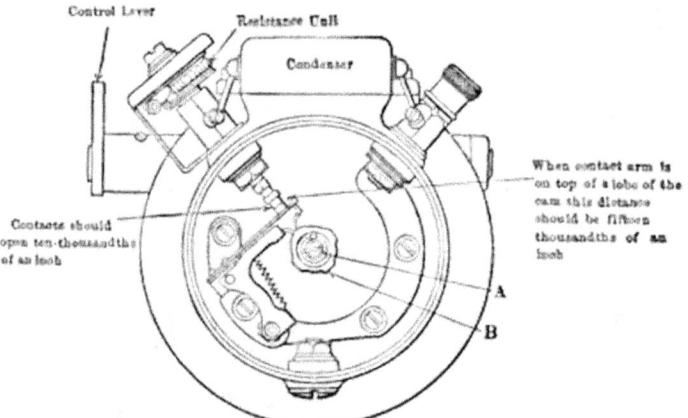

Fig. 166.—The Delco timer.

A weight on the timer shaft acts as a centrifugal governor to operate the automatic spark control. In addition to the automatic spark control a manual control is provided, which is operated by a lever on the steering column, and is connected to the lever at the bottom of the motor generator. The manual spark control is for the purpose of securing the proper ignition control for variable conditions such as starting, differences in gasoline, and weather conditions. The automatic control is for the purpose of securing the proper ignition control necessary for the variations due to speed alone.

The Coil.—The ignition coil is the dark vertical cylinder shown on the front side of the motor generator in Fig. 165. It serves to transform the low voltage current in the primary circuit to a current of high voltage in the secondary circuit. The coil consists of a primary winding of coarse wire wound around an iron core in comparatively few turns, and of a secondary winding of many turns of fine wire, also the necessary insulation and terminals for wiring connections.

106. The Remy-Studebaker Ignition System.—This system, shown in Fig. 167, is built by the Remy Electric Co. and is used on the Studebaker car. It is of the high tension distributor type with the primary current furnished by the storage battery. Dry batteries are supplied for emergency purposes. The storage battery is kept charged by the starting generator. The distributor and breaker box form an individual unit, as shown in Figs. 168 and 169. Figure 169 shows clearly the operation of a distributor, the current entering at the center and being directed by

Fig. 167.—Wiring diagram of Remy-Studebaker ignition system.

the revolving arm to the different contact plates on the inside of the cover. These connect to the different plugs.

The transformer coil is of the non-vibrating type furnishing a single spark, the interruption of the primary circuit taking place in the breaker box. Inside the breaker box is the primary interrupter or circuit breaker. By the action of the cam D the two points A and B close and open twice in each revolution of the shaft. These points are in the circuit of the current flowing from the battery to the primary coil winding. The interruption of this current induces a high tension current in the secondary winding of the coil. The interrupter makes two sparks to one revolution of its shaft and therefore must run at twice the speed of the distributor

Fig. 168.—Face and side views of Remy-Studebaker distributor and breaker box.

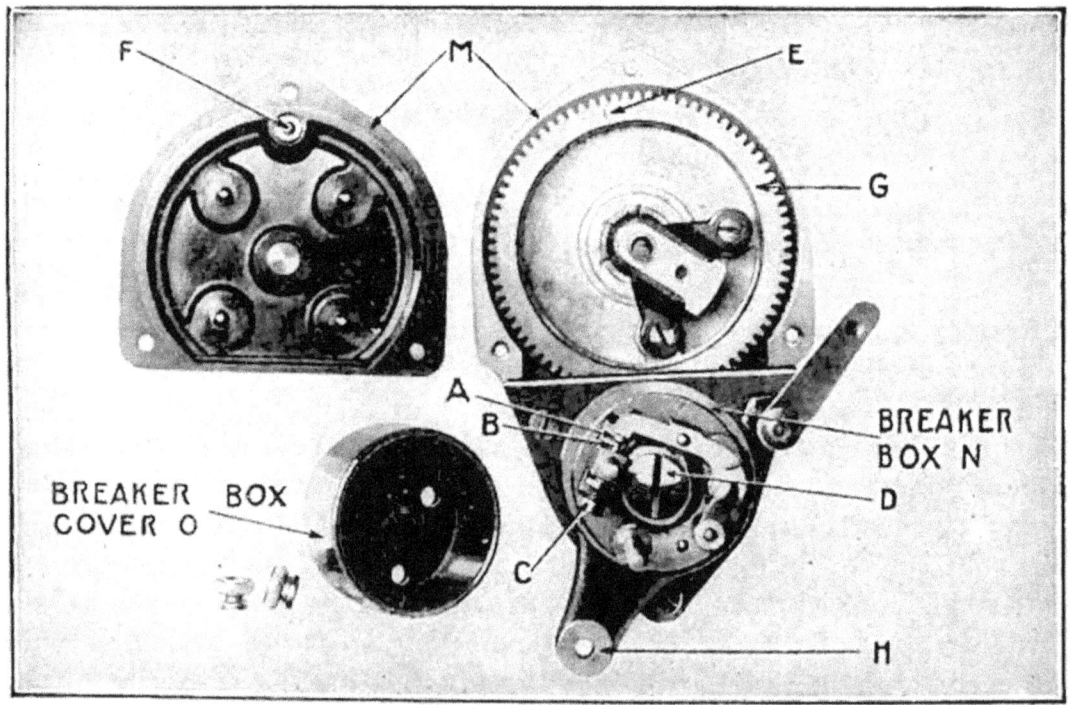

Fig. 169.—Remy-Studebaker igniter disassembled.

BATTERIES AND BATTERY IGNITION

for a four-cylinder engine. For six cylinders it would make three revolutions to one of the distributor.

107. Spark Advance and Retard.—It is very essential in a variable speed gasoline engine that the time at which the spark occurs in the cylinder be changed according to the engine speed, as it takes a certain length of time to produce an explosion, regardless of the engine speed. When the engine speed is high, the spark must occur before the piston reaches dead center in order to have the full force of the explosion when the piston has just passed the center position. When the engine speed is slower, the spark can occur later and yet have the force of the explosion exerted just after dead center. It is necessary when starting that the spark occur not before dead center.

Fig. 170.—Delco automatic spark advance mechanism as used on Cadillac cars.

These various considerations demand that the position of the spark be made variable. This is usually done by shifting the timer, or interrupter housing, causing the break of the primary current (and consequently the spark in the cylinder) to occur earlier or later. The position of the spark is in most cases governed from the steering column. In starting the engine, the spark should not occur until after the piston has started on its down stroke. It should then be advanced as the engine increases its speed. If the spark is too far advanced there will be a decided knock in the cylinders.

108. Automatic Spark Advance.—In several modern ignition systems, means are provided by which the position of the spark is automatically advanced and retarded. This relieves the driver from the responsibility and uncertainty of correctly gauging the position at which to set the spark lever. Figure 170 shows the Delco spark advance mechanism used on the Cadillac. As is seen, it consists of a ring governor which determines just when the timer contact breaks. As the engine speeds up, the ring swings nearer to a horizontal position and this shifts the interrupter cam so that the circuit is broken earlier. A spring pulls them back when the engine slows down. The mechanism of the Atwater Kent automatic spark advance was shown in Fig. 159 and that of the Westinghouse system in Figs. 163 and 164.

CHAPTER VII

MAGNETOS AND MAGNETO IGNITION

109. Principles of Magnetism.—The principle upon which a magneto is constructed involves an understanding of some elementary magnetic and electrical principles in addition to those discussed in the preceding chapter.

Magnets.—It is a well known fact that either in a bar magnet or in a magnet bent in the shape of a horseshoe, as in Fig. 171 the "magnetism," that invisible force which attracts and repels iron or steel, is concentrated near the ends, as indicated by the bunches of iron filings at the ends of these magnets. One end of the magnet is called the "north" or N-pole,

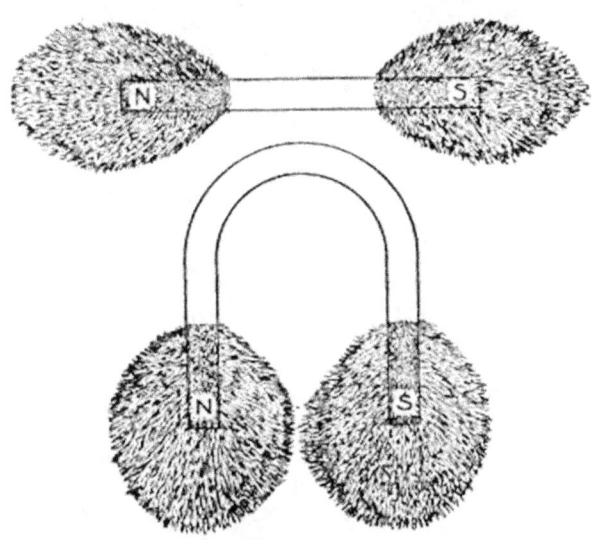

Fig. 171.

and the other the "south" or S-pole. The difference between the two poles can be seen by taking two horseshoe magnets and placing their like poles and again their unlike poles together. It will be found that the "like" poles repel each other and the "unlike" poles attract each other. This is the fundamental law of magnetism.

Lines of Force.—If a horseshoe magnet be placed on its side, as shown in Fig. 172, a piece of paper put over it, and iron filings be sprinkled over the paper, we shall find that the filings arrange themselves in well-defined lines, their direction being as indicated. This arrangement shows us that there is a magnetic force acting between the two poles of the magnet. The direction is shown and, if the investigation be continued, it will be discovered that this invisible force acts from north pole to south pole. These invisible lines are known as magnetic "lines of force."

Permanent and Electro-magnets.—Horseshoe magnets are either "permanent" or "electro" magnets. A *permanent* magnet is one

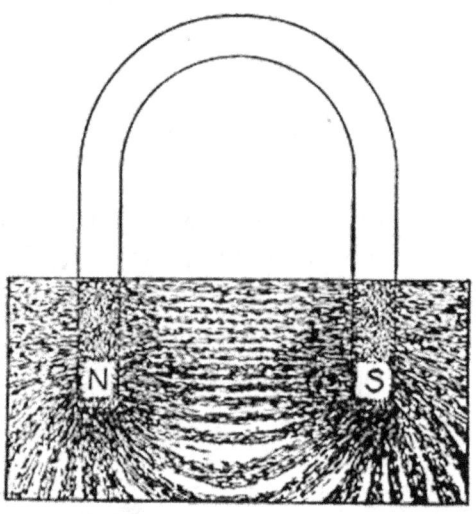

Fig. 172.

made of highly tempered steel which has been magnetized and usually retains its magnetism indefinitely. An *electro-magnet*, Fig. 173, is made of wrought iron or soft steel, and carries a coil of wire through which a current of electricity is passed when the iron or steel is to become magnetized. As soon as the current in the wire is cut off, the magnet loses its magnetism. The name "electro-magnet" signifies that the mag-

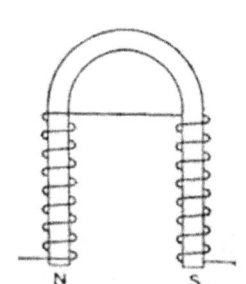
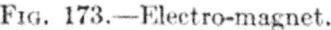
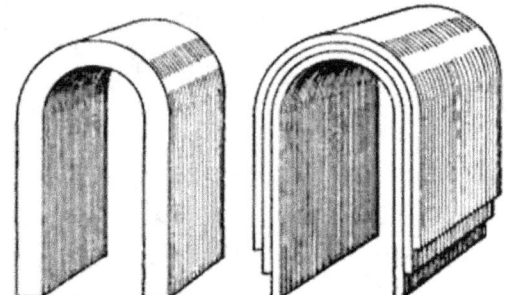

Fig. 173.—Electro-magnet.   Fig. 174.—Simple magnet. Compound magnet.

netism is the effect of the electric current. In the mechanical generation of current we shall see that the magnetism in the horseshoe magnet is made use of. If a permanent magnet is used for creating the magnetic field, the machine is called a "magneto" and if electro-magnets are used, the machine is called an electric generator.

MAGNETOS AND MAGNETO IGNITION

Simple and Compound Magnets.—In some types of magnetos, compound permanent magnets are used. A compound magnet is one built up of several simple magnets, as shown in Fig. 174. It has been found that a compound magnet is much stronger than a simple magnet of the same size.

110. Mechanical Generation of Current.—It is found that if a wire be moved across the magnetic field between the poles of a magnet so as to cut the "lines of force" there will be an electric current generated in the wire. If the wire should now be moved across the lines of force in the opposite direction, the current will also flow in the opposite direction in the wire. The reason for this is not clearly explained, but it is a well known fact that cutting magnetic lines of force by moving a wire across them will generate current in the wire.

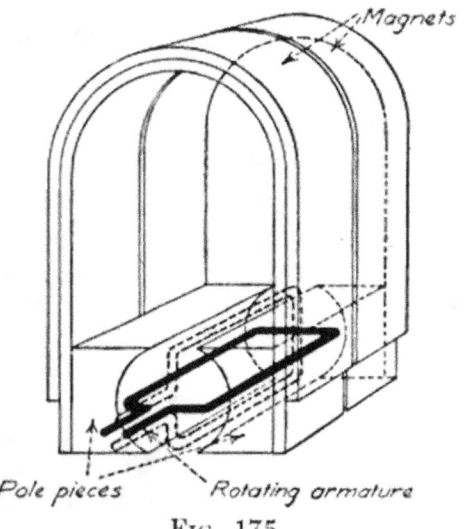

Fig. 175.

This fact is made use of in the magneto, an elementary type of which is shown in Fig. 175. The wire has been formed in the shape of a rectangle and arranged to rotate between the pole pieces of the magnet. If the ends of the wire are connected by a measuring instrument, a current of electricity will be found to flow out of one end of the wire and into the other end as the wire is revolved. This current will be an alternating current; that is, the current changes in direction each time the rectangle

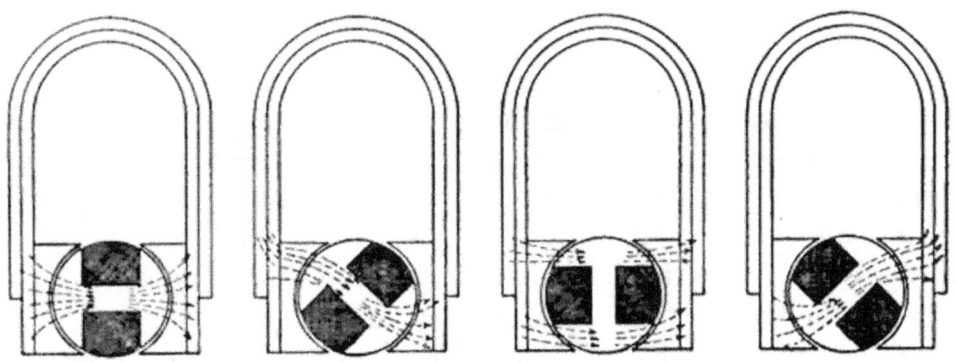

Fig. 176.

turns over. When the wire is cutting the "lines of force" at right angles the voltage is the maximum, and it is at this period of rotation that the current is best for ignition purposes. This condition occurs twice during a complete revolution of the loop of wire.

In an actual magneto, instead of having only one turn of wire, a great many turns of wire are wound in the shape of a coil around a piece of laminated iron, called the armature core. This coil is caused to rotate between the magnetic poles, generating a current in it. Figure 176 illustrates the change and cutting of the magnetic lines of force during one complete revolution of the armature. By using the laminated iron armature core, the flow of the magnetism between the poles of the magnet is increased, thus increasing the lines of force that are cut by the coils of wire.

111. Low and High Tension Magnetos.—A "low tension" type of magneto is one which delivers current of a low voltage, which must be converted to the necessary high voltage for ignition by an external transformer coil. The armature contains only a primary winding, while the transformer coil has the usual primary and secondary windings.

Fig. 177.—Side view—Remy Model P magneto.

A "high tension" magneto delivers current from the armature of sufficiently high voltage for ignition, without the use of an external transformer coil. The high tension current is generated by having two windings on the armature of the magneto, one a primary winding, and the other a secondary winding. The armature assembly also contains a condenser. The true high tension magneto must not be confused with the so-called high tension magnetos in which the armature current is transformed by a coil merely placed in the top of the magneto, instead of outside as is done in the low tension type. The coil is merely contained in the magneto assembly for convenience but this does not make it a "high tension" magneto in the strict sense of the term.

112. Armature and Inductor Types.—An "armature" type of magneto is one in which the lines of force are cut by means of a coil of wire wound on an armature rotating between the magnetic pole pieces, as just described. It may be either of the high or low tension type.

In an "inductor" type of magneto, the coil of wire is stationary. The cutting of the lines of force by the stationary coil is caused by a revolving "inductor." The current is generated in the stationary coil and this avoids the necessity of having sliding contacts and brushes in order to connect the coil with the external circuit. The inductor type may also be "low" or "high" tension. The constructional features of these two general types will be pointed out in considering the several modern magneto types.

FIG. 178.—Distributor end view—Remy Model P magneto.

113. Remy Model P Magneto.—Figures 177 and 178 show side and distributor end views of the Remy Model P magneto, of the low tension armature type.

The Remy armature shown in Fig. 179 is of the H or shuttle type, with laminated core made from soft Norway iron. The armature heads are of hard bronze, and the drive shaft, which is of steel, is cast into the armature head. The armature winding is of cotton covered enameled wire heavily impregnated with a special insulating compound rendering it impervious to heat and moisture. The armature shaft revolves on magneto-type ball bearings which are made dust and grit proof by the use of felt washers.

In a low tension magneto, the current generated in the armature is led through a circuit breaker to the primary winding of the coil. When the circuit breaker is closed, the current flows through the primary winding and magnetizes the core of the coil. At the desired instant for the spark, the circuit breaker opens the circuit quickly and thus destroys the magnetism of the core of the coil. This action induces a high tension current in the secondary winding of the coil. This is led back to the distributor of the magneto, where it is directed to the proper spark plug on the engine.

The armature winding cuts the lines of force twice in each revolution and therefore will give two sparks per revolution. For this reason, there are two lobes on the cam which operates the circuit breaker. For a four-cylinder engine, the magneto armature should run at crank shaft speed, as two sparks are required per revolution of the engine. For a six-cylinder engine, the armature of the magneto should run at one and one-half times crank shaft speed, as three sparks are needed per revolution of the engine. The distributor terminals should be connected to the plugs in the order in which the cylinders are to fire.

The Circuit Breaker.—The circuit breaker illustrated in Fig. 180 may be shifted by the spark lever to change the time of the spark. The breaker points are made of iridium-platinum, which gives them an exceedingly long life. The timing control lever may be located on either side of the magneto, as the circuit breaker and housing are reversible. An ample timing range of 35° is provided for.

Condenser.—The condenser, instead of being placed in the coil, is placed just above the armature. The purpose of the condenser is to prevent sparking at the breaker points, when they break the magneto primary circuit.

Magnets.—The magnets are made from tungsten-steel specially heat treated and hardened, thereby insuring the retention of magnetism for a long period.

Coil.—The coil, the top view of which is seen in the wiring diagram of Fig. 181, has the switch built integral with it. The coil is fastened behind the dash and the switch face only appears on the driving side.

Distributor and Timing Button.—The distributor terminals located on the face of the distributor provide a reliable method of securing the high tension spark plug cables. An ingenious device, known as the timing button, is incorporated in the distributor, for the purpose of timing the magneto to the motor. With this device the circuit breaker and distributor are brought into proper position, thus facilitating this usually difficult operation of timing the magneto to the motor, an operation that frequently puzzles even an experienced repair man.

MAGNETOS AND MAGNETO IGNITION

Fig. 179.—Armature of Remy Model P magneto.

Fig. 180.—Remy Model P magneto—circuit breaker removed.

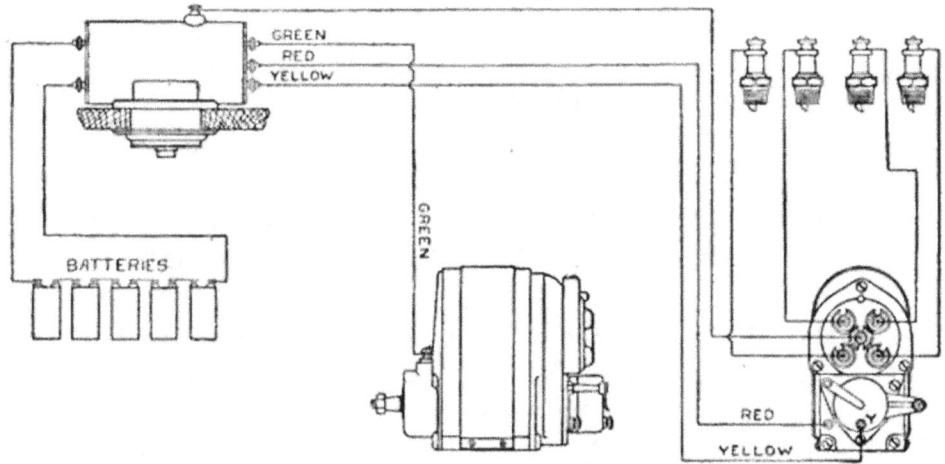

Fig. 181.—Wiring diagram for Remy Model P magneto.

For timing the magneto, turn the engine over by the starting crank until No. 1 piston reaches the top dead center at the end of the compression stroke. Press in on the timing button at the top of the distributor and turn the magneto shaft until the plunger of the timing button is felt to drop into the recess on the distributor gear. This places both distributor and circuit breaker in the proper position corresponding to the engine position given above, and they may now be coupled together.

114. The Connecticut Magneto.—This magneto, illustrated in Fig. 182, likewise has a shuttle wound armature revolving between the poles of permanent magnets, and generates an alternating low tension current with two impulses for each revolution. It has but a single

Fig. 182.—Connecticut magneto partially disassembled.

primary wire running to the switch; all secondary wires connect from the magneto direct to the plugs. The transformer coil is encased in a metal tube in cartridge form and is mounted in the magneto just above the armature.

115. Dual Ignition Systems.—The voltage generated in a magneto depends on its speed, and this makes it desirable to have some other source of current for starting an engine. This auxiliary source is either a set of dry cells or a storage battery. In the dual system the battery supplies the primary current for starting, the current being led through the circuit breaker and primary winding of the coil. On the dual system the regular coil and distributor of the magneto are used. After the engine is started the switch can be thrown to use the magneto current.

MAGNETOS AND MAGNETO IGNITION

116. Eisemann High Tension Dual Ignition.—The wiring diagram for the Eisemann E. M. Dual system is shown in Fig. 183. This magneto is of the high tension armature type. The Eisemann dual system consists of a direct high tension magneto and a combined transformer coil and switch, the transformer being used only in connection with the battery, and the switch being used in common by both battery and magneto systems. The magneto is practically the same as a single ignition high tension instrument. To insure reliability, the vulnerable parts of each system are separate from those of the other. For instance, separate windings and circuit breakers are used for each system. On the other hand, parts

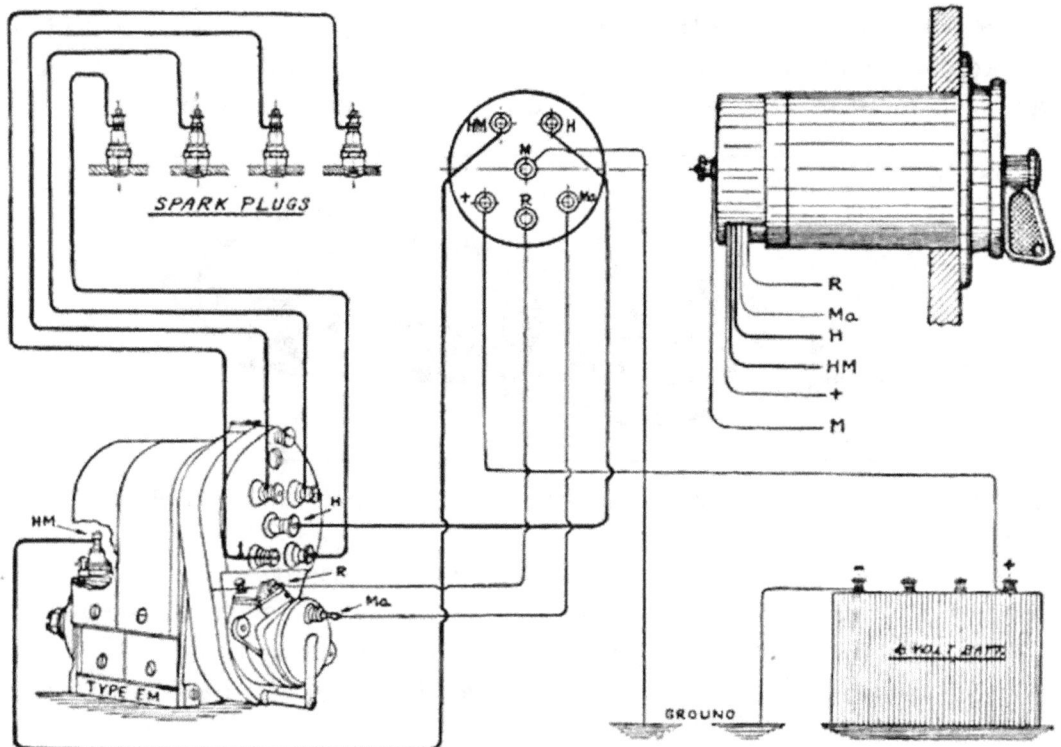

Fig. 183.—Wiring diagram—Eisemann type E. M. Dual four-cylinder ignition system.

that are not subject to accident or rapid wear are used in common, so as to avoid unnecessary duplication.

The magneto armature is an iron core, made of many pieces of soft sheet iron riveted together, around which is a primary winding of medium-gauge copper wire. Over this primary winding, is a secondary winding consisting of many coils of very fine copper wire, the wire being specially insulated in the entire length and the layers being carefully insulated from each other. The low tension current, formed by rotating the armature, in turn induces a secondary or high tension current in the secondary winding. The transformation of the low tension current into high tension current is obtained by suddenly interrupting the low tension current by the circuit breaker or make-and-break mechanism. It will thus be seen that the high tension armature is practically a transformer coil wound directly on the armature core with a circuit breaker to interrupt the primary current.

Spark Control.—As the spark occurs when the primary circuit is broken by the opening of the platinum contacts, the timing of the spark can, therefore, be controlled by having these platinum contacts open sooner or later. This latter is accomplished by the angular movement of the timing lever body. This movement gives a timing range of 30°. The spark is fully retarded when the timing lever is pushed as far as possible in the direction of rotation of the armature and is advanced when pushed in the opposite direction.

Safety Spark Gap.—If a spark plug cable becomes disconnected or broken, or should the gap in the spark plug be too great, then the secondary current has no path open to it and, in endeavoring to find a circuit,

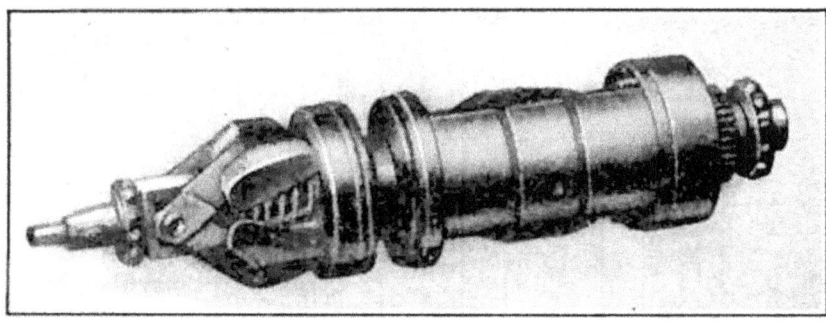

Fig. 184.—Eisemann armature with automatic spark advance mechanism.

will sometimes puncture the insulation of the armature or of the coil. To obviate this, a so-called "safety spark gap" is placed on the top of the armature dust cover. It consists of projections of brass with a gap between them. One of these is an integral part of the dust cover, and therefore forms a ground.

The Coil.—The coil of Fig. 183 is designated as Type D C and consists of a non-vibrating transformer and a switch which is used in common to put either the battery or magneto ignition into operation. The coil is cylindrical in shape, is compact, and is placed through the dashboard. The end which projects through on the same side as the motor has terminal connections for the tables. The other end, facing the operator, contains the switch and the starting mechanism. The transformer coil is used only in conjunction with the battery. There is a push button circuit breaker in the center of the switch for producing a spark with the battery current when the engine is not running. The coil is provided with a lock and key, so that the switch may be locked in the "off" position.

MAGNETOS AND MAGNETO IGNITION

117. Eisemann Automatic Spark Control.—The automatic spark control magneto is of the same construction as the standard high-tension instrument with the addition of the automatic mechanism as shown in Fig. 184. The automatic advance is accomplished by the action of centrifugal force on a pair of weights attached to one end of a spiral sleeve between the shaft of the magneto and the armature. When the armature is rotated, the weights begin to spread and exert a longitudinal pull on the sleeve, which in turn changes the position of the armature with reference to the pole pieces. In this way, the moment of greatest induction is advanced or retarded and with it the break in the primary circuit. The cams which lift the circuit breaker and cause the break in the primary circuit are fixed in the correct position with relation to the armature, so that the break occurs at the moment when the current in the winding is strongest.

118. The K-W High Tension Magneto.—The K-W high tension magneto is of the alternating current inductor type. Figure 185 is an external view and Fig. 186 shows a longitudinal sectional elevation. By referring to the numbers, an idea can be obtained of the function of the various parts.

64 Driving pinion.	1 Bridge.
79 Plunger for primary circuit.	100 High tension lead.
67 Cam.	96 Distributor block.
68 Cam roller.	73 Magnets.
189 Retainer spring.	130 Rotor.
56 Switch binding post.	114 Primary winding.
98 Distributor brush holder.	113 Secondary winding.
120 Secondary contact plunger.	126 Condenser.
119 Secondary distributor plunger.	118 Safety spark gap.
	186 High tension bus bar.
2 Distributor gear.	14 Low tension bus bar.
	10 Base.

The only revolving part in the K-W magneto is shown in Fig. 187. This part is the rotor which is constructed of fine laminations of the softest Norway sheet iron. These laminations are riveted together, are accurately bored out to fit the rotor shaft, and are accurately machined as to width and diameter, being mounted on this shaft at exactly right angles to each other. Between these two pieces is the stationary winding or coils, also shown separately in Fig. 188. The winding, which is concentric with the armature shaft, is mounted in between the two halves of the rotor and stands absolutely still. In the position shown, the lines of force go straight across through the right hand rotor. When the shaft turns 45° from this position, the rotors connect the magnetism from one pole piece, through the center of the winding, to the opposite pole piece, thus giving a powerful wave of current from a quarter revolution of the magneto.

The winding, shown in Fig. 188, is a double winding, that is, it has a primary or low tension winding, which is surrounded by a secondary or high tension winding. This primary winding goes to the circuit breaker of the magneto, where its current is interrupted when the spark is

Fig. 185.—K-W high tension magneto.

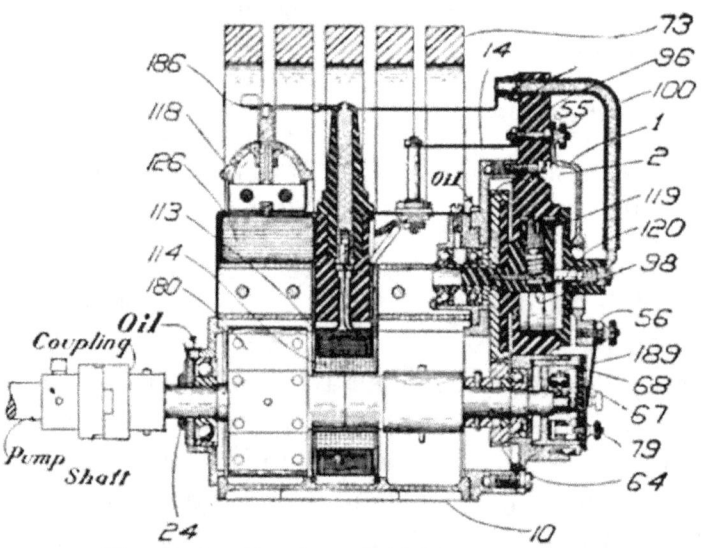

Fig. 186.—Section of K-W magneto.

wanted and during one of the periods of armature rotation in which considerable current is generated.

At the moment of this interruption of current in the primary, a powerful surge of current is generated in the secondary winding. The current from this secondary winding goes straight up through the hard rubber terminal to the high tension bus bar, as shown in Fig. 186, to the center of the distributing brush and from there is distributed to the various cylinders of the motor.

The condenser, No. 126, Fig. 186, is bridged across the circuit breaker points. Its function is to absorb the low tension current after the

Fig. 187.—K-W rotor and coils.   Fig. 188.—K-W coils.

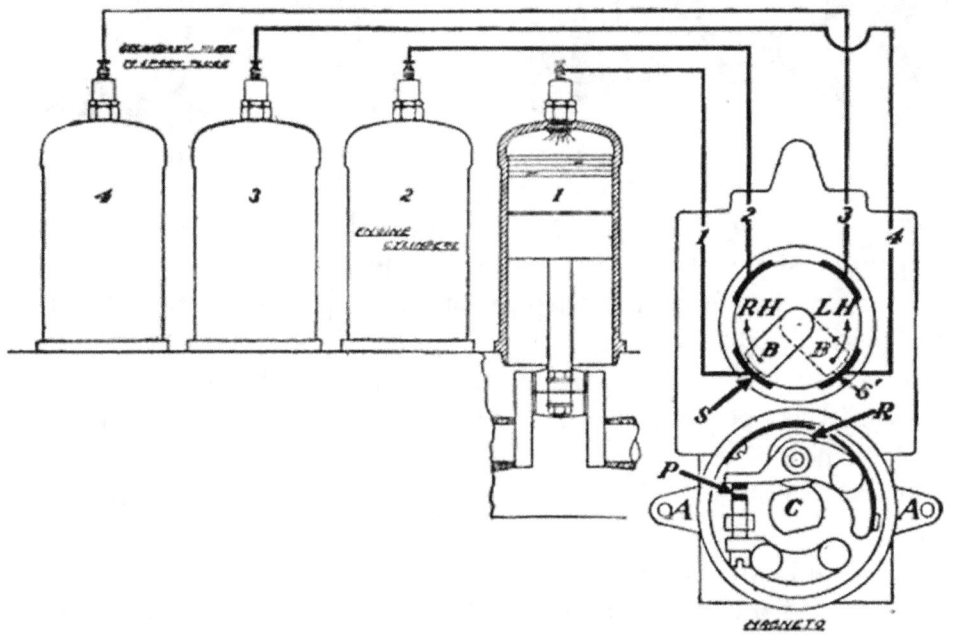

Fig. 189.—Wiring diagram for K-W high tension magneto.

breaking of the primary circuit at the breaker points. This condenser is made of a large number of sheets of tinfoil and mica.

The safety gap, No. 118, Fig. 186, is a necessary part of any high tension magneto, its object being to form a path for the high tension current to jump through in case a secondary cable that leads to the spark plugs should be off when the engine is running. This safety gap, as its name implies, prevents the magneto from burning out, for as long as there is a path for the high tension current to pass through, it will never puncture the insulation of the secondary winding.

It will be noted by referring to Fig. 186 that the distributor shaft is carried on two ball bearings, as is also the rotor shaft. The distributor block is moulded from a special composition of hard rubber, and is accurately machined all over. The brass segments that connect with the various plug holes on top of the distributor are moulded into the hard

Fig. 190.—Dixie magnets and rotor. Fig. 191.—Dixie coil and field pieces.

rubber. A carbon brush is mounted in the distributor arm, which presses slightly against the distributor segments, and the interior of the distributor is practically dust and moisture proof, being protected by a hard rubber cover, held in place by a three-legged spider or bridge, No. 1. This bridge also carries the primary circuit to the circuit breaker. The binding post, No. 56, is the point from which the switch wire is run to the switch for the purpose of cutting out the circuit breaker and stopping the engine.

Figure 189 is a wiring diagram for the K-W high tension magneto, Type H.

119. The Dixie Magneto.—The Dixie magneto is built upon a principle different from that of either the armature or the inductor types. Figure 190 indicates the arrangement of the magnets and the rotating element carried in bearings by the two pole pieces. This rotor turns between the pole pieces and, as the iron pieces simply form extensions to the magnet pole pieces and are always of the same polarity, there is no reversal of magnetism through them.

Just above the rotor, and with its axis at right angles, is placed the coil, supported by the two upright field pieces enclosing the armature as shown in Fig. 191. Figures 192, 193, 194, and 195 show the reversal of the

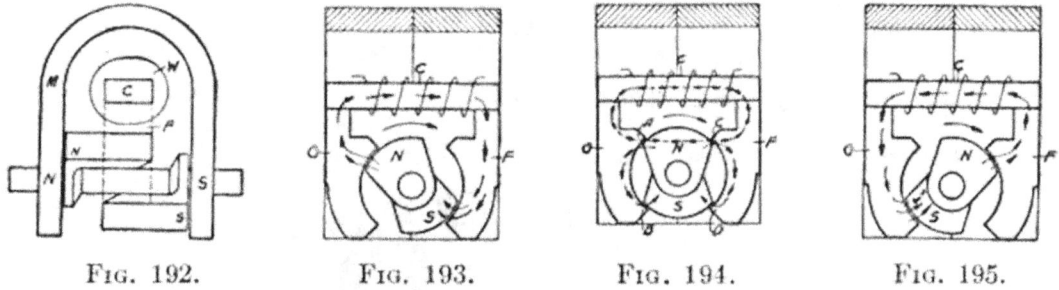

Fig. 192.   Fig. 193.   Fig. 194.   Fig. 195.

Figs. 192 to 195.—Showing the principle of the Dixie magneto.

lines of force through the coil during one-half revolution of the rotor. This change of the lines of force through the coil, which has a primary and a secondary winding, causes a low tension alternating current in the primary winding, and this induces the high tension current in the secondary winding when the contact points break the primary circuit. Figure 196 is a diagrammatic sketch of the primary circuit. P is the primary coil, A

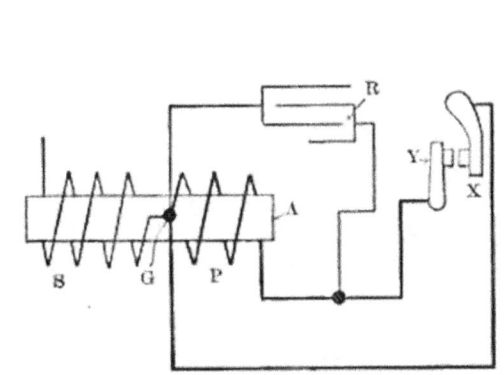

Fig. 196.—Primary circuit of Dixie magneto.

Fig. 197.—Bosch high tension magneto.

is the core, R is the condenser, X and Y are the circuit breaker points, G is the common ground connection for both primary and secondary windings, and S is the secondary coil.

120. The Bosch High Tension Magneto.—The Bosch magneto, shown in Fig. 197, is of the high tension armature type, generating two sparks during each revolution of the armature shaft. A longitudinal section of a Bosch magneto is shown in Fig. 198 and a rear view in Fig. 199. The principal numbered parts are as follows:

- 1 Brass plate at the end of the primary winding.
- 2 Fastening screw for contact breaker.
- 119 Long platinum contact screw.
- 118 Short platinum contact screw.
- 9 Condenser.
- 120 Lock nut for contact screw 119.
- 121 Flat spring for magneto interrupter lever.
- 105 Holding spring for interrupter cover.
- 10 High tension collector ring.
- 11 Carbon brush for high tension current.
- 12 Holder for brush.
- 13 Fastening nut for brush holder.

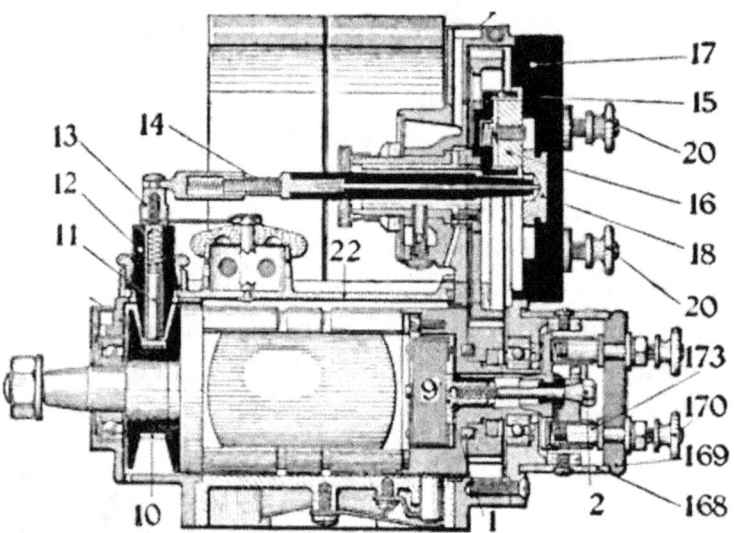

Fig. 198.—Section of Bosch high tension magneto.

- 14 Spring contact for conducting the high tension current.
- 15 Distributor brush holder.
- 16 Distributor carbon brush.
- 17 Distributor disc.
- 18 Central distributor segment.
- 20 High tension terminals.
- 22 Dust cover.
- 123 Interrupter lever.
- 168 Interrupter housing and timing lever.
- 169 Cover for interrupter housing.
- 173. Low tension brush.

The beginning of the primary winding is grounded to the armature core and the other end is connected to the brass plate 1. In the center of this plate is the fastening screw 2, which serves first, for holding the contact breaker in its place, and second, for conducting the primary current to the platinum screw block of the contact breaker. Screw 2 is insulated from the contact breaker disc, which is in metallic connection with the armature core. The platinum screw 119 is fixed in the contact piece and receives the current from screw 2. Pressed against this platinum screw, by means of the spring shown, is the magneto interrupter lever 123 with platinum screw 118, which is connected to the armature core and, therefore, with the grounded end of the primary winding. The primary circuit is, therefore, closed as long as the magneto interrupter lever 123 is in contact with platinum screw 119. The circuit is interrupted when the lever is rocked by the cam so as to open the contact. The condenser 9 is connected across the gap formed when the contacts break.

The beginning of the secondary winding is connected to the insulated end of the primary so that the one forms a continuation of the other. The other end of the secondary winding leads to the collector ring 10,

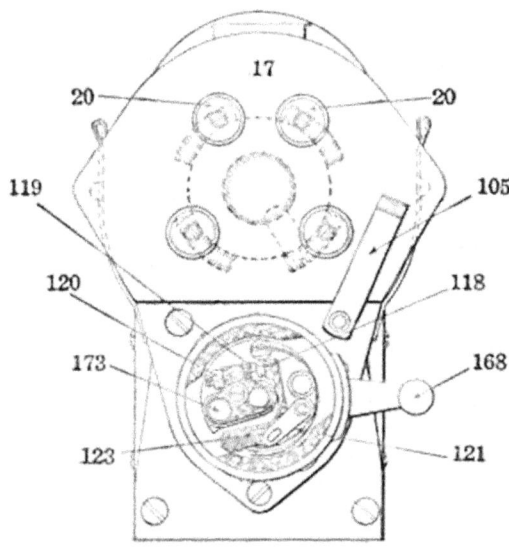

Fig. 199.—End view of Bosch high tension magneto.

on which slides a carbon brush 11, held by the carbon holder 12, and thus insulated from the magneto frame. From the brush 11 the secondary current is conducted to the terminal 13, through the spring connection 14 to the center distributor contact 18, and from there to the carbon brush 16, the latter rotating with the distributor gear wheel.

In the distributor disc 17, metal segments are embedded, and as the carbon brush 16 rotates, it makes contact with the respective segments of the distributor. Attached to the metal segments of the distributor are the connection terminals 20 to which are fixed the conducting cables to the spark plugs.

From the end of the secondary winding the high tension current is distributed to the respective cylinders in the order in which they operate. The current produces the spark which causes the explosion; it then returns through the motor frame and the armature core back to the beginning of the secondary winding. The diagram of connections is shown in Fig. 200.

Safety Spark Gap.—In order to protect the insulation of the armature and of the current conducting parts of the apparatus against excessive voltage, a safety spark gap is provided as shown in Fig. 200. The current will pass through this gap in case a cable is taken off while the magneto is in operation or if the electrodes on the spark plugs are too far apart. The discharges, however, should not be allowed to pass through the safety gap for any length of time; special care has to be taken in this respect if the motor is equipped with a second system of ignition, in

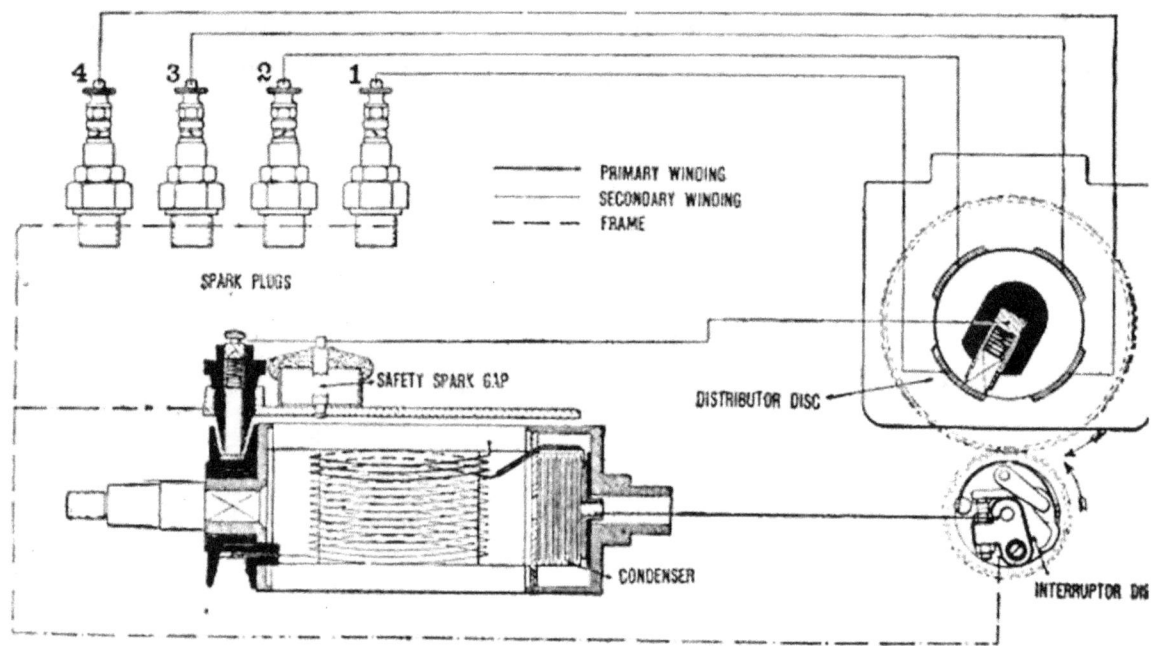

Fig. 200.—Wiring diagram of Bosch high tension magneto.

which case it is necessary to short circuit the primary winding, as the continued discharge of the current over the safety gap is likely to damage the magneto.

121. The Bosch Dual System.—In the Bosch dual ignition system, the standard type of Bosch magneto is used with the application of two timers or interrupters. The parts of the regular current interrupter are carried on a disc that is attached to the armature and revolves with it, the rollers or segments that serve as cams being supported on the interrupter housing. In addition, the magneto is provided with a steel cam which is built into the interrupter disc and has two projections. This cam acts on a lever supported by the interrupter housing, the lever being so connected in the battery circuit that it serves as a timer to control the flow of battery current. These parts may be seen in Fig.

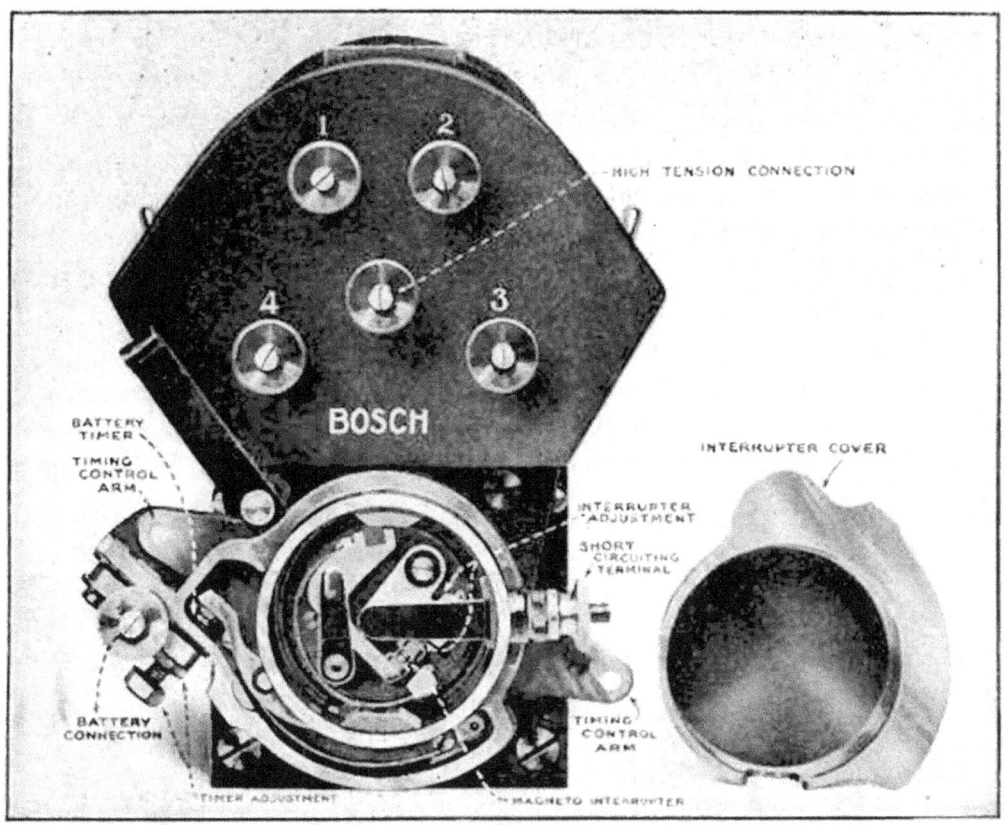

Fig. 201.—Bosch dual system, showing magneto interrupter and battery timer.

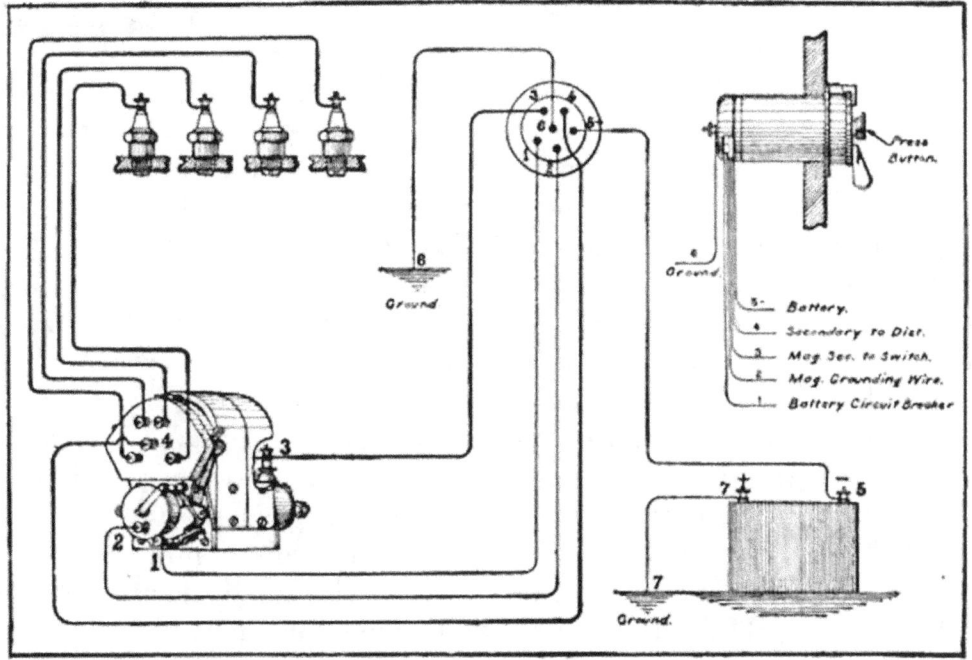

Fig. 202.—Wiring diagram for Bosch dual system.

201. A non-vibrating transformer coil is used with the battery current to produce the necessary voltage.

It is obvious that the sparking current from the battery and from the magneto can not be led to the spark plugs at the same time, so a further change from the magneto of the independent form is found in the removal of the direct connection between the collecting ring and the distributor. The collecting ring brush shown in Fig. 198 as No. 11 and in Fig. 202 as No. 3, is instead, connected to the switch, and a second wire leads from the switch to the central terminal on the distributor. When running on the magneto, the sparking current that is induced in the secondary armature winding flows to the distributor by way of the switch contacts. When running on the battery, the primary circuit of the magneto is grounded, and there is, therefore, no production of sparking current by the magneto; it is then the sparking current from the

Fig. 203.—Parts of Bosch dual coil.

coil that flows to the central distributor connection. It will thus be seen that of the magneto and battery circuits the only parts used in common are the distributor and the spark plugs.

The Bosch Dual Coil.—The Bosch dual coil used in the dual system consists of a cylindrical housing bearing a brass casting, the flange of which serves to attach the coil to a dashboard or other part. The coil is provided with a key and lock, by which the switch may be locked when in the "Off" position. This is a point of great advantage, for it makes it unlikely that the switch will be left thrown to the battery position when the engine is brought to a stop. The absence of such an attachment is responsible in a large measure for the accidental running down of the battery. This locking device also prevents the unauthorized operation of the engine. The parts of the coil are shown in Fig. 203. In addition to the housing and end plate, they consist of the coil itself, the stationary switch plate, and the connection protector.

When the engine is running on battery ignition, a single contact spark is secured at the instant when the battery interrupter breaks its circuit, and the intensity of this spark permits efficient operation of the engine on the battery system.

Starting on the Spark.—For the purpose of starting on the spark, a vibrator may be cut into the coil circuit by turning the button that is seen on the coil body in Figs. 202 and 203. Normally, this vibrator is out of circuit, but the turning of the button places it in the battery primary circuit instead of the circuit breaker on the magneto. A vibrator spark of high frequency is thus produced.

It will be found that the distributor on the magneto is then in such a position that this vibrator spark is produced at the spark plug of the cylinder that is performing the power stroke; if mixture is present in this cylinder, ignition will result and the engine will start.

Connections.—In the wiring diagram of this system as shown in Fig. 202, it will be noted that while the independent magneto requires but one switch wire in addition to the cables between the distributor and spark plugs, the dual system requires four connections between the magneto and the switch; two of these are high tension and consist of wire No. 3 by which the high tension current from the magneto is led to the switch contact, and wire No. 4 by which the high tension current from either magneto or coil goes to the distributor. Wire No. 1 is low tension, and conducts the battery current from the primary winding of the coil to the battery interrupter. Low tension wire No. 2 is the grounding wire by which the primary circuit of the magneto is grounded when the switch is thrown to the off or to the battery position. Wire No. 5 leads from the negative terminal of the battery to the coil, and the positive terminal of the battery is grounded by wire No. 7; a second ground wire No. 6 is connected to the coil terminal.

122. Bosch Two-independent System.—The Bosch two-independent or double system consists of two complete and independent systems of ignition. One consists of a Bosch high tension magneto system and the other of a Bosch high tension distributor battery system.

The battery system is utilized for starting purposes and for emergency ignition in case of accident to the magneto system, which is used for ordinary service. The battery system consists of a combined coil and switch and a timer-distributor, which are completely independent of the magneto. The two systems are brought together at the switch, and the connections are such that the engine may be operated on the magneto with one set of plugs, or on the battery with the other set of plugs, or on the magneto and battery together, in which case both sets of plugs are used. Either the battery or magneto may be used for ignition with the other system entirely dismantled or removed from the engine. The wiring diagram for this system is shown in Fig. 204.

123. The Ford Magneto and Ignition System.—The magneto which generates the current for the ignition system in the Ford car is of the low tension alternating current type and differs from the conventional type in that the stationary and revolving elements are interchanged.

The Ford magneto, as shown in Fig. 205, has but two parts, a stationary armature, consisting of a number of coils, which are attached to a stationary support in the flywheel housing, and a set of permanent field magnets of the horseshoe type, which are secured to the flywheel, the whole being a part of the motor. The magnets revolve with the flywheel at a distance of $\frac{1}{32}$ in. from the coils, in which the current is

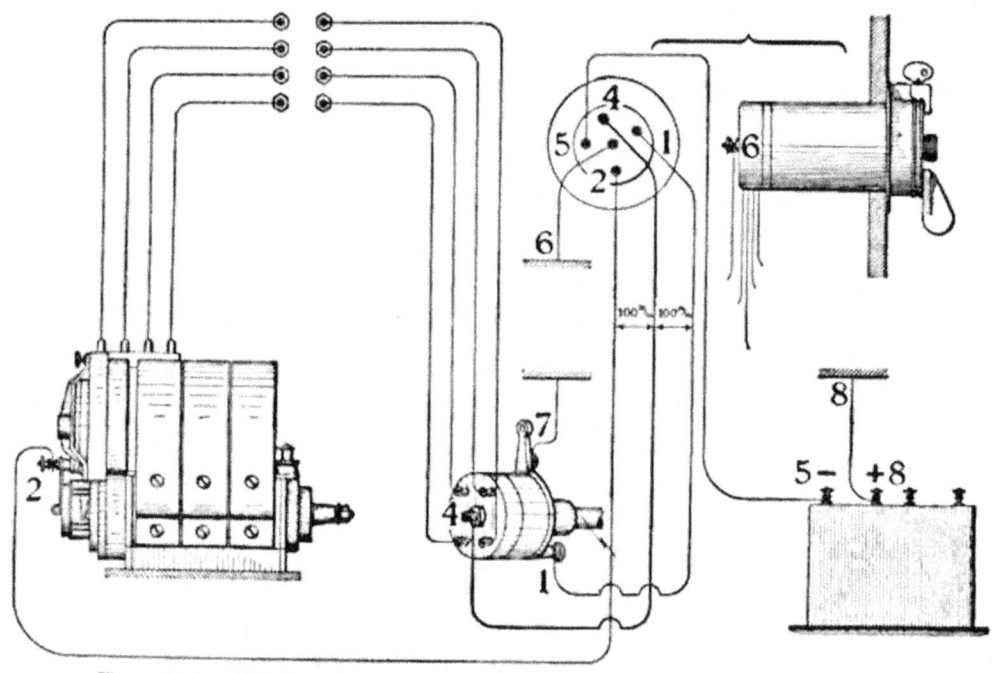

FIG. 204.—Wiring diagram for Bosch two-independent system.

induced by the magnetic field. The current flows to the four spark coils, passing through whichever one is at the instant connected to the ground by the commutator. The coils are the ordinary double winding vibrator coils. The induced current from each coil goes to its spark plug to perform its function of igniting the charge. The magneto and its component parts are fully illustrated in Fig. 206.

The diagram of Fig. 207 shows the plan of wiring of the Ford Model T motor, which, it will be noted, is very simple. The current generated by the magneto flows through the primary winding of the coil whose circuit is closed by the commutator, to the commutator, and back through the frame of the motor to the magneto. This completes the primary circuit or path of the magneto current. The high tension induced in the secondary winding of the coils is led to the spark plugs in the cylinders as their respective primary circuits are completed by the commutator.

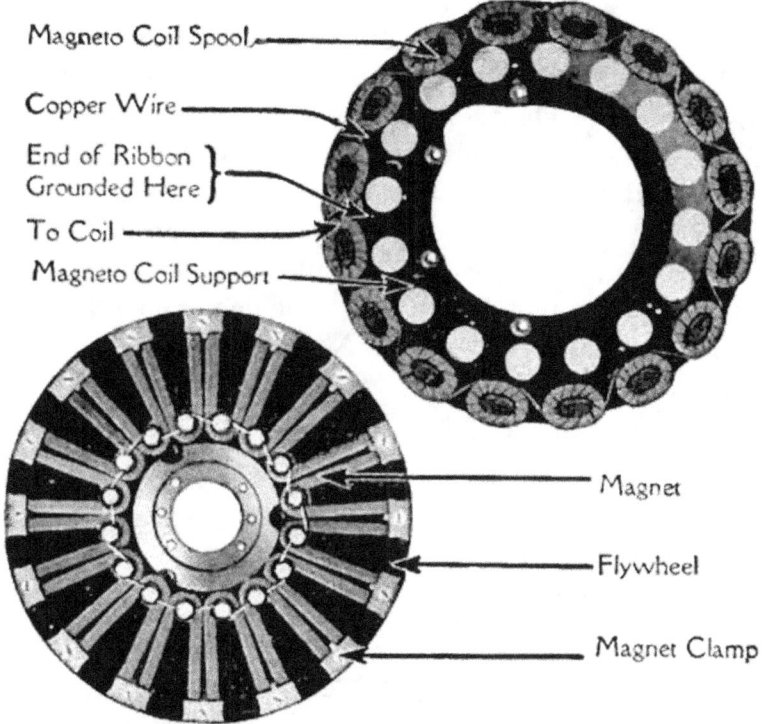

Fig. 205.—The Ford magneto.

124. Magneto Speeds.—Nearly all of the modern magnetos are constructed, as was pointed out in Art. 113, page 157, to give one spark for each one-half revolution of the armature or inductor. This means that

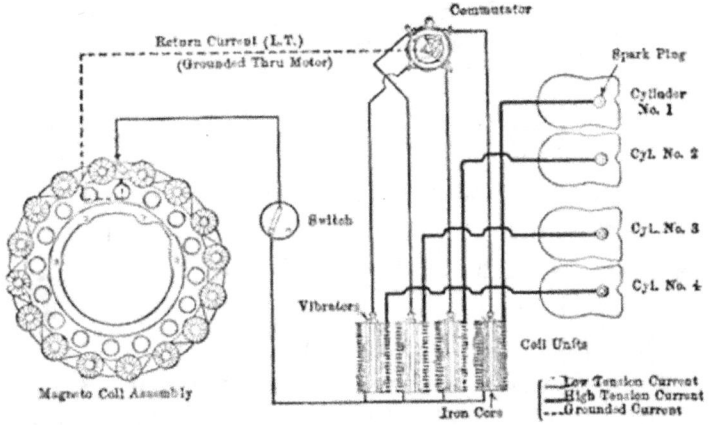

Fig. 206.—Diagram showing the course of circuit through the Ford ignition circuit.

for each revolution, two sparks are obtained from the magneto. For a four-cylinder four-stroke engine, there are two explosions per revolution of the crank shaft. We see, therefore, that the magneto and engine crank shaft must run at the same speed. For a six-cylinder four-stroke engine, there are three explosions per revolution of the crank shaft, requiring one and one-half revolutions of the magneto. The magneto must, therefore, run one and one-half times the crank shaft speed. Some magnetos are built to give four sparks per revolution. These must, of course, be set to run at one-half the speeds given above.

125. Timing the Magneto.—Necessarily, the rules for setting and timing magnetos must be very general. If the magneto has been removed or is out of adjustment, the engine should be cranked until the No. 1 piston (the one next the radiator) is on dead center at the end of the compression stroke. This position can usually be found by markings on the flywheel. On some engines the manufacturers recommend that the engine be cranked just a few degrees past the dead center. The position will then be the firing position for the No. 1 cylinder.

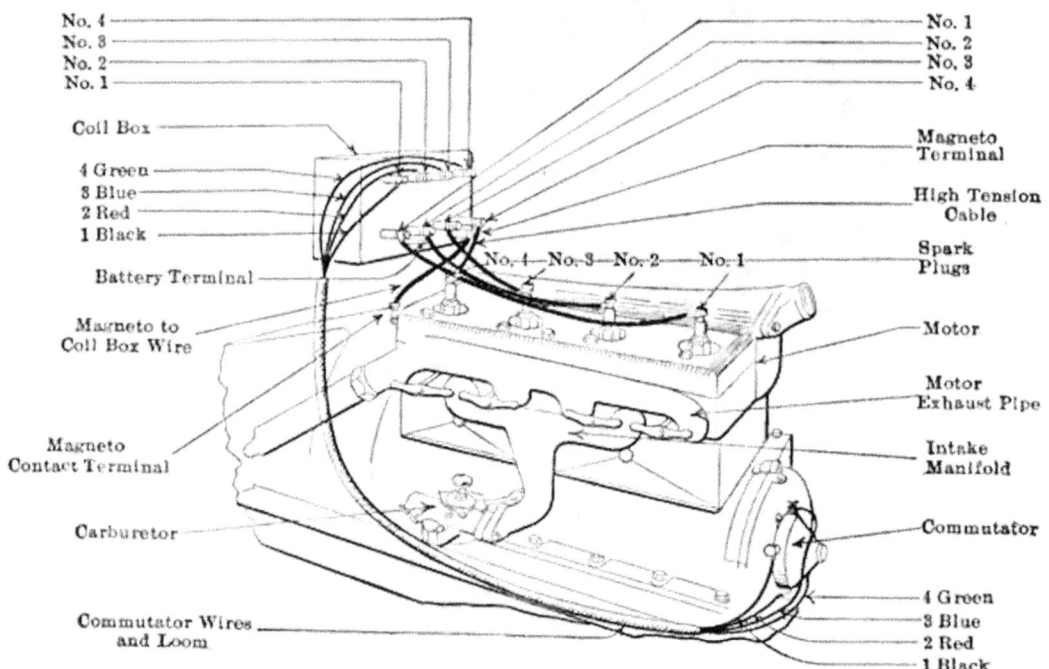

Fig. 207.—Wiring of the Ford ignition system.

The distributor housing should then be taken off and access gained to the distributor mechanism. It should also be determined just which cylinder corresponds to each of the distributor points. The armature should then be rotated until the distributor segment comes in contact with the distributor point for No. 1 cylinder. Adjust the armature so that the contact points just break when the interrupter housing is in full retard and attach it to the driving shaft. The spark control rod should now be connected and adjusted so that the contact points just open, when the spark lever on the steering wheel is in full retard. This permits the maximum spark advance.

126. Battery vs. Magneto Ignition.—It is a somewhat common idea that an engine will run faster on a magneto spark than on a battery spark. This contention has been frequently advanced in support of magneto ignition. Extensive experiments on engines equipped with a double system, one a magneto and the other a battery system, prove that with the same spark setting, there is practically no variation in engine speed, provided both systems are in perfect order and adjustment. In individual cases where the contrary has been found it was probably due to some weakness or defect in the system which was replaced and should not be taken as condemning that type of ignition in general.

127. General Suggestions on Magnetos.—The magneto should never be tested unless the whole system is completely assembled with all parts and wires in place and attached. Water should be kept away from all parts of the ignition system. Magnetos were not intended to be run in water.

Care should be taken when oiling parts of the magneto. A small amount of oil properly placed is essential, but a great lot on everything is a constant source of trouble.

Don't take the magneto apart or try to improve its construction. Repairing a magneto is an expert's work. Unless you are one, don't attempt it.

128. Common Magneto Ignition Definitions.—*Low Tension Magneto.*—A magneto which generates a low voltage current, requiring a transformer coil to raise the voltage for ignition purposes. Only one winding is found on the armature.

High Tension Magneto.—One which generates current of high enough voltage for ignition purposes. The armature contains two windings, a primary and a secondary winding. No outside coil is necessary.

Armature Type Magneto.—One in which the current is generated by a coil of wire wound around a core revolving between the poles of a permanent magnet.

Inductor Type Magneto.—A type of magneto in which the coil is stationary and the lines of force through the coil are changed in direction by means of a rotating inductor.

Dual System of Ignition.—A system of ignition with two sources of current, magneto and battery, either of which may be used. There is practically no duplication of equipment, as the magneto timer, distributor and plugs are used for both sources of current.

Double System of Ignition.—Two complete systems of ignition with nothing in common excepting the switch on the dashboard. There is a duplication of practically the entire equipment, plugs, timer, and distributor.

CHAPTER VIII

STARTING AND LIGHTING SYSTEMS

129. Starting on the Spark.—If an engine is stopped with an explosive mixture in the cylinder, it may sometimes be started from rest by merely causing a spark in the cylinder. In a four-stroke engine having four or more cylinders there will always be one cylinder on the expansion stroke and one on the compression stroke. On a four-cylinder automobile we can sometimes swing the spark lever so as to cause a spark in one of these cylinders, and, if the compression has not been lost entirely, or the gasoline vapor has not been condensed, the engine will start. Sometimes an engine can be started in this manner after standing for several hours. To make an engine more sure of starting on the spark, the throttle should be opened wide before the engine is stopped. This will insure a good charge in each cylinder. When a four-cylinder motor comes to rest after the spark is shut off, one piston will be on its exhaust stroke and another will be on its suction stroke, both of these cylinders, therefore, being open to the air. A third piston will be on its compression stroke with all valves shut and the fourth will be going down on the expansion stroke with its charge still fresh because the current has been turned off. The motor will come to rest with these two pistons on the same level, each about halfway in the stroke. To start the motor, turn the switch to the battery side and press the ignition starter button. Pressing the ignition starter button short-circuits or cuts out the timer or circuit breaker and causes current to flow through the primary winding of the coil. Releasing the push button breaks the primary circuit and causes a high tension current in the secondary circuit, which will be conducted to a spark plug provided the distributor arm is opposite one of the distributor segments.

If the engine comes to rest with the piston which is on the working stroke on the same level with the piston which is on the compression stroke, the distributor arm will be nearer to the segment leading to the cylinder whose piston is on the working stroke. If the spark occurs in this cylinder the engine will be run in the desired direction and if the explosion is sufficient to carry the next piston over the top of the compression stroke, the regular cycles will be continued; but if, when the engine stops, the pistons have gone beyond the position where they are on the same level, the spark is apt to occur in the cylinder which is on the compression stroke. This explosion will drive the engine backward. Near the end of this backward stroke the inlet valve will open and the burnt gases will be discharged through the carburetor.

If the engine is stopped so that the timer points or circuit breaker points are in contact, it is impossible to start by pressing the ignition starter button, but starting may be accomplished by retarding the spark control lever and opening and closing the ignition switch, several times if necessary.

The same method of starting will apply to two- or three-cylinder, two-stroke engines. If a two-stroke engine is started by *advancing* the spark, the motor will continue to run, but in the opposite direction from that desired. A common way of starting a single-cylinder two-stroke engine is to retard the spark and then turn the engine backward by hand until the spark occurs. The engine will then be propelled in the desired direction.

The failure of engines to start on the spark after standing for some time is largely due to the gasoline vapor being heavier than air. After an engine has stood for some time the heavy vapor will settle, and, if the engine is cold, the gasoline may condense on the piston and cylinder walls.

130. Mechanical Starters.—Self starters may be divided into four general types: mechanical starters, air starters, acetylene starters, and electric starters.

Mechanical starters include the various types of hand cranking devices and springs. The disadvantage of the hand cranking starter is that it requires a certain amount of human power. The only advantage is that the driver does not have to leave his seat to crank the engine. The spring starter is capable of giving the engine a few revolutions only, and if the engine does not start then, it becomes necessary for the driver to wind up the spring, which is a rather tiresome operation. If the motor starts, there is an automatic device by which the spring is wound up by the engine.

131. Air Starters.—In the air starters, the air is pumped into a storage tank at about 150 lb. pressure. The engine is started by admitting air into the combustion chamber. The pipe leading from the tank goes to a distributor which is driven by the motor. In this way the air gets only to the cylinder which is on the working stroke and has all the valves closed. This system has the disadvantage that the air is liable to cool the cylinder and prevent proper starting of the regular cycle on account of the gas condensing on the cool walls.

132. Acetylene Starters.—Some manufacturers have equipped their machines with a device for starting with acetylene gas. This gas is very explosive and will ignite readily under almost any conditions.

These engines are equipped with valves and tubes from the acetylene lighting system so that the driver can inject a small quantity of acetylene gas into the cylinders. The engine will then be practically sure of starting on the spark. This system has been largely superseded by the electric starter.

133. Electric Starters.—A still further development in this line is the electric starter. Electric starters may be divided into three types: first, the single-unit system; second, the two-unit system; and third, the three-unit system. In the first system the motor-generator unit furnishes the current to charge the storage battery and operate the lights, and also acts as a motor in cranking the engine. The two-unit system has a generator for charging the battery and furnishing the current for lighting and ignition, but it has a separate unit (a direct current motor) for cranking the engine. The three-unit system has a generator used solely for charging the battery and operating the lights, a motor for cranking the engine and a magneto for furnishing current for ignition.

In all electric self-starters it is necessary to have a storage battery to store up the current so that there is a ready source of sufficient current to drive the motor for starting. The units of the self-starting system are: the generator to furnish electricity; the storage battery which acts as a reservoir to hold the supply of current; and an electric motor to crank the engine. The electric starter may be directly connected to the gas engine, or it may be driven by a set of gears, or by a silent chain.

In order that the electric motor will not be overspeeded when the engine picks up, it is necessary to have an overrunning clutch. This device operates only when the engine runs faster than the motor. The reduction in gears between the electric motor and the engine is about 25 to 1, which means that the electric motor must run twenty-five times as fast as the engine. If it were not for the over-running clutch, the electric motor would be driven at excessively high speed, when the engine picks up to, say, two or three hundred revolutions per minute. The over-running clutch is automatic. It permits the electric motor to drive the engine, but breaks the driving connection as soon as the engine speeds up to a higher rate than the motor is running at. In the one-unit and two-unit systems, the current for ignition is taken from the storage battery. In all cases the current for the lights comes from the battery when the engine is running at low speeds.

There is also another type of self starter which takes the place of the engine flywheel. This unit is a motor-generator outfit and has no reduction gear whatever.

134. Storage Batteries.—A commercial storage cell, as shown in Fig. 208, is made up of the following parts: a jar or container usually made of rubber, positive and negative plates, separators between the plates, and the electrolyte. The electrolyte is a solution of sulphuric acid and water. After the plates are prepared, they are placed in the container and the electrolyte added. The current is then passed through the plates and solution. In this manner the battery is charged. When the battery is fully charged, the electrolyte or solution in the cells should have a specific gravity of 1.27 to 1.29. The specific gravity will become lower as the battery discharges and, when completely discharged, should not be lower than 1.15 to 1.17, or about twelve points less than when fully charged. Water must be added occasionally to replace the loss by evaporation. If one cell regularly requires more water than the others, it is an indication of a leaky jar. A leaky jar should be immediately replaced by a new one. The specific gravity of the electrolyte is the

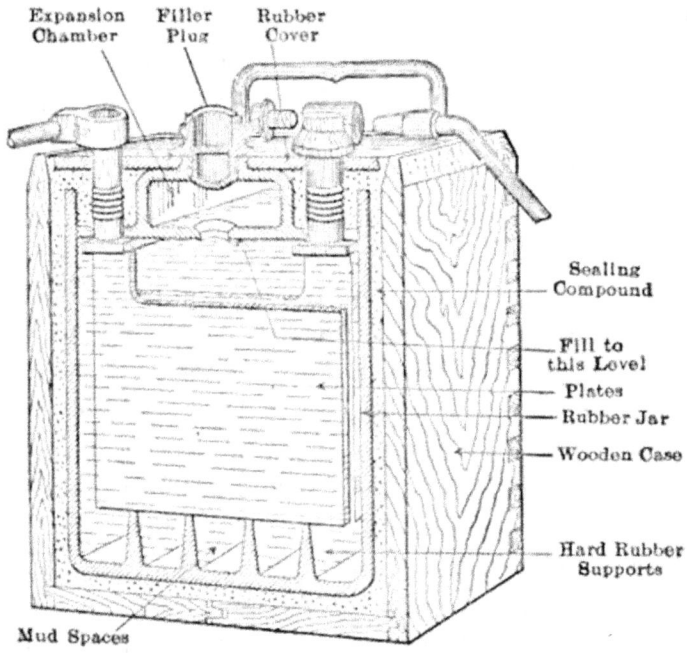

Fig. 208.—Section of storage cell.

most reliable indication of the state of charge of the battery. It should never go below 1.15, for below that the battery will not have sufficient power to turn over the engine and it will not burn the lights so as to give the full candle power. The electrolyte must always cover the plates.

The loss by evaporation should be replaced by adding pure fresh water. The water for filling the batteries must be either distilled water, melted artificial ice, or fresh rain water. Never add acid. The batteries should be inspected once every 2 weeks and, if the electrolyte is below the bottom of the filling tubes, water enough should be added to bring the level up to the proper point. Ordinarily it will require only a few spoonfuls. The filling plugs must be replaced and screwed up tight after filling. Never keep the supply of water in a metal container, a bucket or can. It is best to get a bottle or jug of distilled water from your druggist or from the ice plant. The main point is to keep metal particles out of the battery. Spring water, well water, or hydrant water from iron pipes will contain iron and other materials in solution which will cause trouble by short circuiting the plates.

If the electrolyte has been spilled from a cell, replace the loss with new electrolyte and follow with an overcharge, either by running the engine for several hours, or by charging from an outside source. In replacing electrolyte, have the specific gravity the same as in one of the adjacent cells. This can be determined by use of the hydrometer. When new electrolyte is required, either to replace loss from spilling, or when removing the sediment, or to replace a broken jar, it can be made by mixing chemically pure sulphuric acid, having a specific gravity of 1.84, and distilled water in the proportions of 1 part of acid to 3 parts of water, by volume. The acid should always be poured into the water, and not the water into the acid. A glass, or other acid-proof vessel, thoroughly cleaned, should be used for mixing the electrolyte. When replacing the cell, be sure that the positive and negative connections have the same positions as before. Then apply vaseline or grease to the studs and nuts before making the connections.

After standing for some time, sediment will accumulate in the bottom of the jar. This should always be removed before it reaches the bottom of the plates. It can be determined by inspection, and will be indicated by lack of capacity, excessive evaporation and overheating when charging. If the battery needs repairing, it is best to communicate with the manufacturers who will advise you what to do. The battery is the heart or center of the system. The electricity generated by the dynamo is stored in the battery, and is used by the starting motor to crank the engine, and for the lights at low speed and when the engine is at rest. When the current flows from the dynamo through the battery elements, it is termed *charging*, and when the battery is supplying current for cranking the engine or to the lights, it is termed *discharging*.

Immediately upon receipt of a battery or new automobile, the battery should be inspected. Remove the vent plugs. See that the battery plates are well covered with solution, and if it is not up to the inside cover (see Fig. 208) add distilled water. Filling one cell does not fill all the cells. The battery, if neglected, will cause the entire system to fail. The starting motor may operate when the battery is weak, but the battery life is thereby shortened. If, however, the battery is kept fully charged, and properly supplied with pure water, it will give uninterrupted service. The majority of car owners are careless about giving the battery the attention it should have. Remember that if the plates are exposed (not covered by battery solution) they become sulphated and hard, and the battery capacity is greatly reduced.

Specific gravity tests are made with the hydrometer. When the battery does not give the desired results, specific gravity tests of each cell will indicate the faulty cell or cells. Figure 209 shows the ordinary type of hydrometer syringe used in determining the specific gravities of solutions.

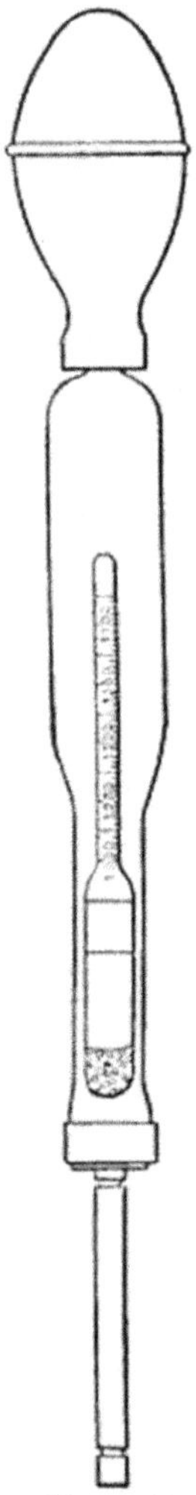

The action of this hydrometer is similar to that shown in Fig. 95, but it is contained in a syringe by which a sample of electrolyte may be drawn from the cell. To use the hydrometer expel the air from bulb by pressing it. Then insert the nozzle into the battery opening and allow the depressed bulb to draw sufficient electrolyte into the syringe to float the hydrometer. The specific gravity or density of the electrolyte is then indicated by the number on the hydrometer stem at the surface of the electrolyte. Always return the battery solution to the cell from which it was taken.

Take the hydrometer readings just previous to adding water. If the hydrometer readings show that one cell is discharged, or nearly so, while the other cells are charged, it indicates that the cell is defective. This may be due to:

1. Short circuits in that particular cell, thus discharging it.
2. Sulphating of the plates, caused by infrequent filling with water or by allowing to stand discharged.
3. Leak in the cell, thereby requiring more water than other cells, which reduces the gravity.

Freezing of the electrolyte is avoided by keeping the battery fully charged. As the specific gravity of the electrolyte decreases (result of discharging), freezing will occur at temperatures as follows:

FIG. 209.—Hydrometer syringe.

FREEZING POINTS OF ELECTROLYTE

Specific gravity	Condition of charge	Freezing point
1.285–1.300	Fully charged.	Can not freeze.
1.260	¼ discharged.	50° below zero.
1.210	½ discharged.	20° below zero.
1.160	¾ discharged.	0° zero.
1.120 or lower.	Discharged.	20°–30° above zero.

While it is possible to freeze a fully charged battery, it can be done only by very low artificial temperatures.

If battery is allowed to remain discharged or if plates are not well covered, the elements become sulphated, and the capacity is thereby reduced. Sulphate can sometimes be removed by a prolonged low charging rate, but more frequently the battery is beyond redemption. The plates should always be well covered and needless discharge prevented.

If the starting motor is used unnecessarily for cranking the engine or for propelling the car, rapid discharge takes place. Avoid this whenever possible, as under this condition the dynamo must be operated a long time to replace in the battery the amount of current taken by the starting motor.

If the battery is neglected the center and upper portions of plates become sulphated. This condition is not due to any fault of battery material or construction nor to the starting-lighting units, but is directly attributable to inattention and neglect on the part of the car owner who has failed to add sufficient distilled water to the solution in each cell, in order to keep the plates properly submerged. Be sure to add distilled water to the battery every week or two.

135. Battery Charging.—Figure 210 shows how lamps are connected in a direct current circuit for battery charging. Connect a wire *A* from one side of lighting source to one side of these lamps, and to other side connect another wire *B*. Then connect wire *C* to the other side of lighting source. When the other end of this third wire *C* is connected to the end of wire *B*, the lamps should light. Now determine which is the positive

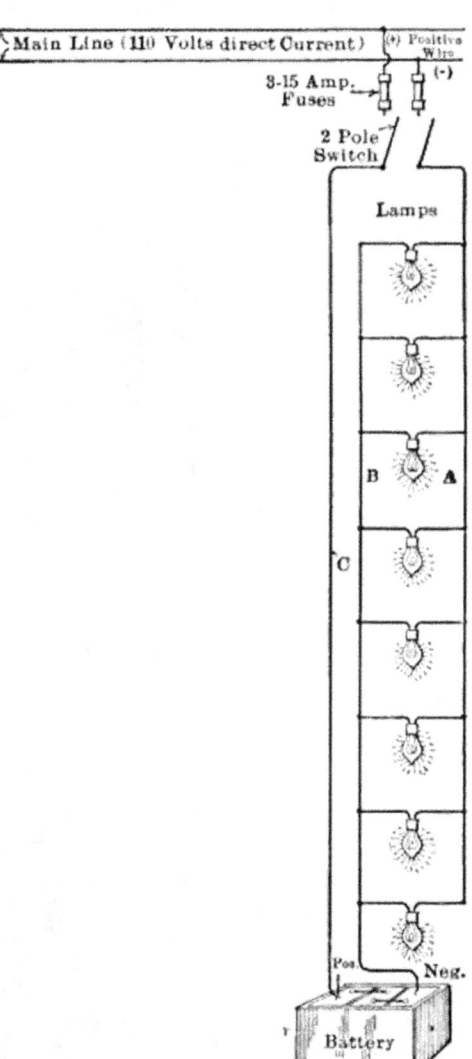

Fig. 210.—Direct current charging method.

(+) and which is the negative (−) wire. Disconnect these two wires *C* and *B* which caused lamps to light, and dip the ends in a bowl of water containing a few tablespoonfuls of salt or one tablespoonful of battery solution. Hold the immersed ends about ¼ in. apart. The wire from which the small bubbles rise is the negative (−) wire. This wire should be connected to the negative battery post, marked *Neg.* or (−). The other wire, which is positive (+) should be connected to the positive

186   THE GASOLINE AUTOMOBILE

battery post marked *Pos.* or (+), but not until the proper amount of resistance has been determined.

If the direct current is at 110 volts any of the following sets of lamps can be used as a resistance to permit a current of 4 amp. to flow into the battery to charge it:

> 8–110 volt, 16 c.p. (50 watt) carbon lamps.
> 4–110 volt, 32 c.p. (100 watt) carbon lamps.
> 16–110 volt, 25 watt mazda or tungsten lamps.
> 7–110 volt, 60 watt mazda or tungsten lamps.

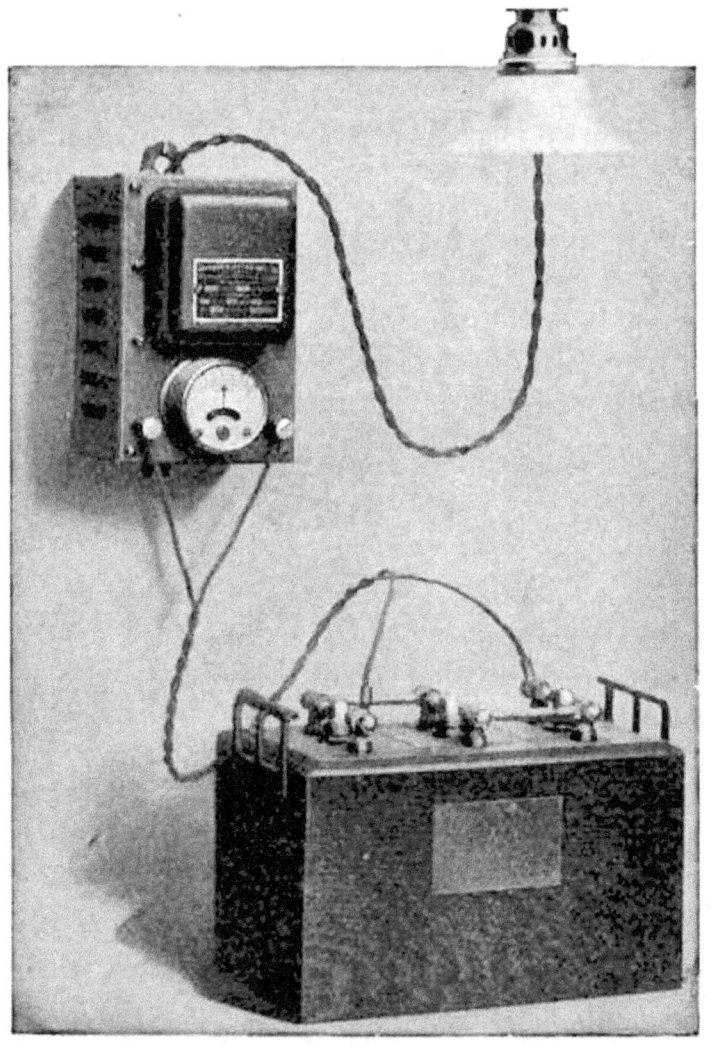

Fig. 211.—The Wagner rectifier charging a storage battery.

The charging operation should continue for 24 to 30 hours, or for two periods of 15 hours each.

If the voltage or pressure is 220 volts, use sixteen 220-volt lamps of 16 c.p. each, or eight lamps of 32 c.p. each; and charge for 24 to 30 hours.

If only alternating current is available the batteries can be charged by a rectifier (see Fig. 211) which can be procured through an electrical supply house. A rectifier is an electrical device for changing alternating to direct current. In ordering, state the voltage and frequency of the line from which the charging current is to be taken. The ordinary lighting circuit has a voltage of 110 and a frequency of 60, but it is best to get this information from the electric light company. In addition to this, the voltage and capacity of the battery must be given. To charge the battery through a rectifier, connect the rectifier in the line, as shown in Fig. 211, following the directions accompanying the instrument.

136. Wiring Systems.—Electric starting systems may be of the single wire or the two wire system. In the two wire system, each unit, such as lamps, motor, and coil, has two wires running to the battery. In the single wire system, one side of the battery is grounded, that is, one wire is bolted to the frame of the car, and each unit has only a one wire connection. In this method it is necessary to have some sort of cut-out, so that if the single wire should become grounded to any metallic part, it would not injure the battery. Any ground on the single wire system would, of course, short-circuit the battery. The cut-out will allow only a certain amount of current to flow and anything in excess of this will cause sufficient magnetism in the core of the cut-out to break the circuit. This action can be detected by a clicking noise, similar to the working of a telegraph instrument.

Fig. 212.—Ward-Leonard controller.

There are a large number of starting and lighting systems on the market, the details of which we will now take up. The most important technical features to study are the different methods of controlling the output of the dynamo.

137. The Ward-Leonard System.—The Ward-Leonard constant current type of controller is shown in Fig. 212 and operates as follows: The proper charging of the battery is automatically regulated by the controller. When the car speed becomes approximately 7 miles per hour, the dynamo armature will give a voltage sufficient to charge the batteries. The circuit between the dynamo and the batteries is normally open, but when the voltage of the dynamo becomes proper for charging, the coil A on the magnet core B magnetizes the core sufficiently to attract the arm C. This arm moves toward the core B, and thus two spark-proof points D and D' are brought together, establishing the circuit between the battery and the dynamo, and the dynamo begins to charge the batteries.

Unless some method of controlling it is adopted, the dynamo voltage increases with the speed. The dynamo should charge at about 7 miles per hour, but when the car runs at a much higher speed, as 15 to 60 miles per hour, it is desirable that the dynamo voltage shall not increase. If allowed to increase, such an excessive dynamo voltage would tend to cause sparking at the brushes, excess current and consequent trouble at the commutator, and excessive wear and heating of the bear-

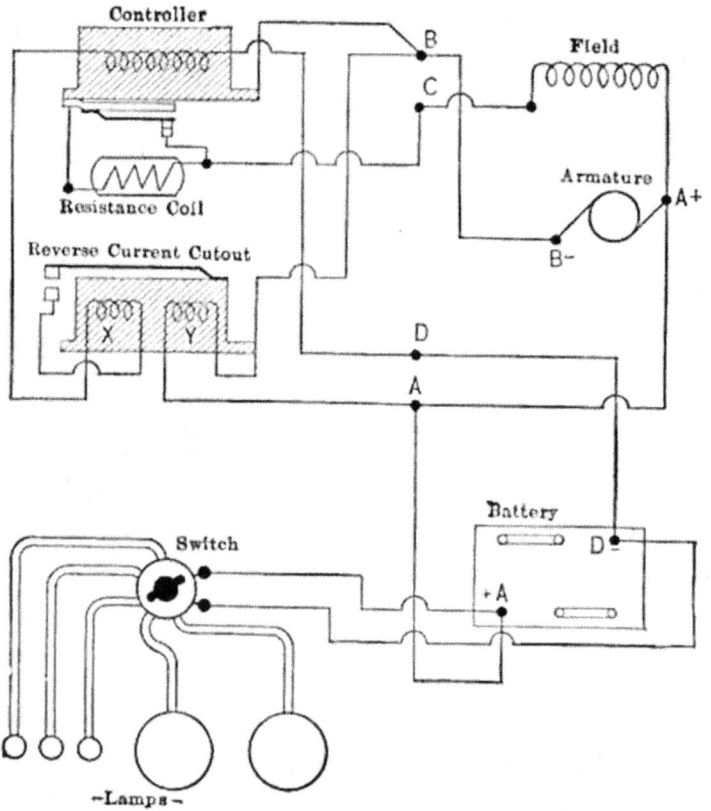

Fig. 213.—Ward-Leonard wiring diagram.

ings. It would also cause an excessive amount of current to flow through the battery. To prevent this, the strength of the dynamo field, and consequently the output of the dynamo, is made dependent on the touching of the two points E and E'. The coil F on the magnet core G carries the armature current, and if this current becomes a certain amount (usually in practice 10 amp.) the core becomes sufficiently magnetized to attract the finger H. This separates the contacts E and E', and a resistance M is inserted in the field circuit. This weakens the fields and causes the amperes to decrease. When the current decreases to a predetermined amount (say 9 amp.), the coil F does not magnetize the core G enough to overcome the pull of the spring J. The spring pulls together the points E and E'; the full field strength is restored and the current tends to increase. Under operating conditions the finger H automatically and rapidly vibrates at such a rate as to keep the current constant. As a result, the dynamo will never charge above a predetermined amount (10 amp.) no matter how high the speed of the car, but will produce a substantially constant current.

In case the engine speed becomes so low that the dynamo volts are less than those of the battery, the magnetism caused by the coil A, Fig. 212, is weakened so that the spring K pulls the contacts D and D' apart. Thus, the circuit between the dynamo and battery is opened

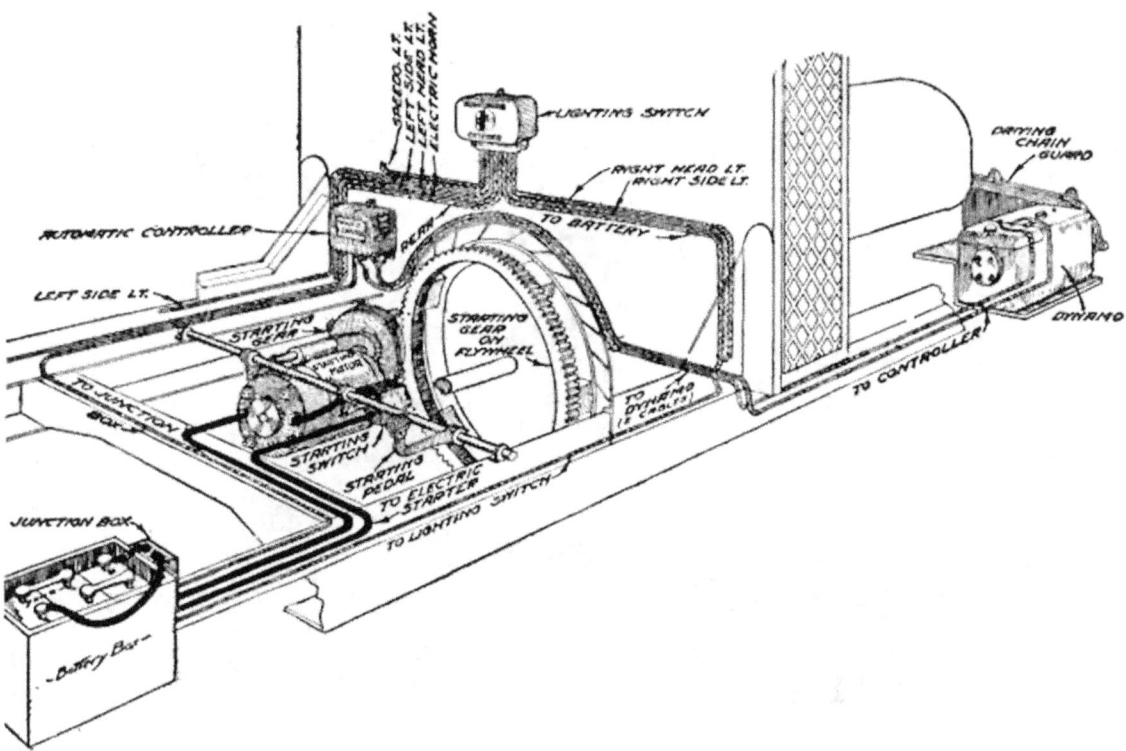

Fig. 214.—Installation of Ward-Leonard system.

when the dynamo speed is too low for proper charging. An auxiliary series coil L on core B acts to insure the perfect demagnetization of the core on reversal of current.

The technical internal wiring diagram, in Fig. 213, shows the connections of the dynamo, the battery and the controller. Figure 214 shows a typical installation and wiring layout for the complete two-unit starting and lighting system. The connections of the motor are very simple. There are two wires from the battery to the motor, with a switch operated by the foot pedal. This pedal also shifts the starting gears into mesh with the teeth on the flywheel. When the engine starts, the foot pedal is released, the gears are disengaged, the switch opened and the motor becomes inoperative until it is wanted to start the engine again.

138. The Delco System.—A single-unit motor-generator is used in this system. This unit also carries, mounted on it, the ignition system. A general view of a Delco system is shown in Fig. 215. The motor-generator has separate sets of brushes, commutator, and windings—one used when serving as a motor and one when acting as a generator. It also has two driving connections. When acting as a generator, it is usually driven from the pump shaft by a clutch connection as shown at the right in Fig. 216, which shows the motor-generator as used on the 1915 Buick cars. When the starting pedal is operated, this clutch is disconnected, the gear

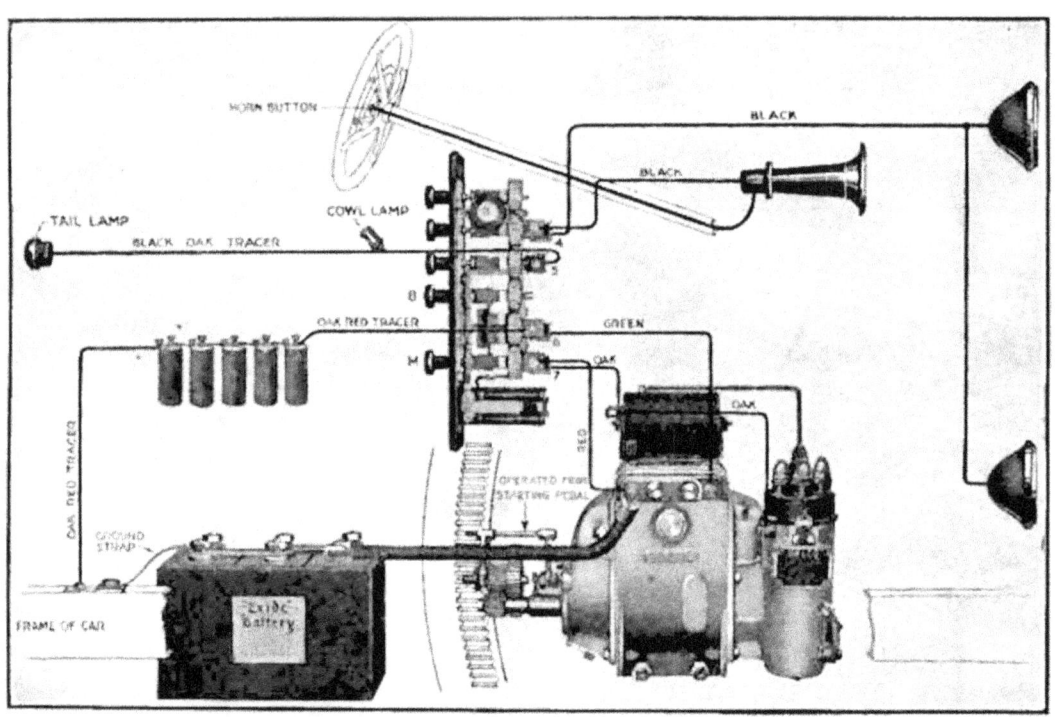

Fig. 215.—Delco starting and lighting system.

connection is made from the motor pinion to the flywheel, the brushes are removed from the generator commutator and the motor brushes put into contact with the motor commutator. When the pedal is *released*, the connections are made to operate as a generator.

Voltage Regulator.—The Delco system of current regulation uses a resistance coil immersed in a tube of mercury, as shown in Fig. 217. This instrument serves to control the amount of current flowing from the generator to the storage battery. By referring to Fig. 217 the construction and operation of this device will be made clear. A magnet coil *A* surrounds the upper half of the mercury tube *B*. Within this mercury tube is a plunger *C*, comprising an iron tube with a coil of resistance wire wrapped around the lower portion on top of a special insulation. One end of this resistance wire is connected to the lower end of the tube, the other end being connected to a needle *D* carried in the center of the plunger. The lower portion of the mercury tube is divided by an insulating tube into two concentric wells, the plunger tube being partly immersed in the outer well, and the needle in the inner well. The space in the mercury tube above the body of mercury is filled with an especially treated oil which serves to protect the mercury from oxidization, to lubricate the plunger, and to form a dash pot for the plunger. Inasmuch as the voltage of the storage battery varies with its condition of charge, the intensity of the magnetic pull exerted by the magnet coil *A* upon the

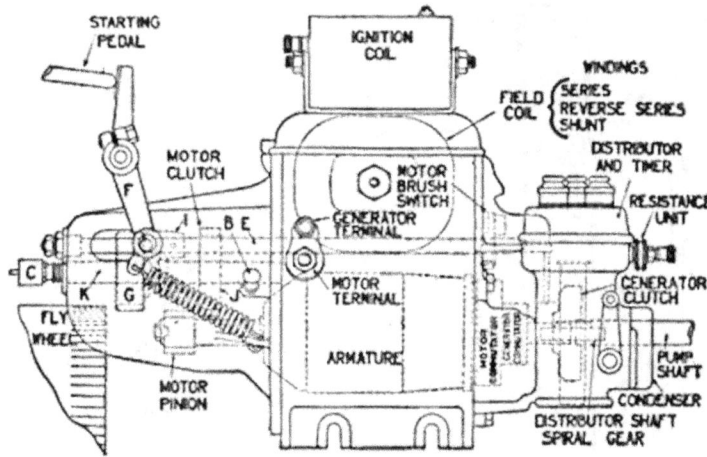

Fig. 216.—Delco motor-generator.

plunger *C* varies, and causes the plunger to move in and out of the mercury as the voltage changes. When the battery is in a discharged condition, the plunger *C* assumes a low position in the mercury tube. When the plunger is at a low position, the coil of resistance wire carried upon its lower portion is immersed in the mercury, and as the plunger rises the coil is withdrawn. Now the current to the shunt field of the generator must follow a path leading to the outer well of mercury, through the resistance coil wound on the plunger tube, to the needle carried at the center of the plunger, into the center well of mercury and out of the regulator.

It will be seen that, as the plunger is withdrawn from the mercury, more resistance is thrown into this circuit, due to the fact that the current must pass through a greater length of resistance wire. This greater resistance in the field of the generator causes the amount of current flowing to the battery to be gradually reduced as the battery nears a state of complete charge, until finally the plunger is almost completely withdrawn from the mercury, throwing the entire length of resistance coil into the shunt field circuit, thus causing a condition of practical electric balance between the battery and generator, and obviating any possibility of overcharging the battery.

Automatic Cut-out Relay.—The automatic cut-out, Fig. 218, is located between the voltage regulator and ignition relay, in the apparatus box. This instrument closes the circuit between the generator and the storage battery when the generator voltage is high enough to charge the storage battery. It also opens the circuit as the generator slows down and its voltage becomes less than that of the storage battery, thus preventing the battery from discharging back through the generator. The cut-out

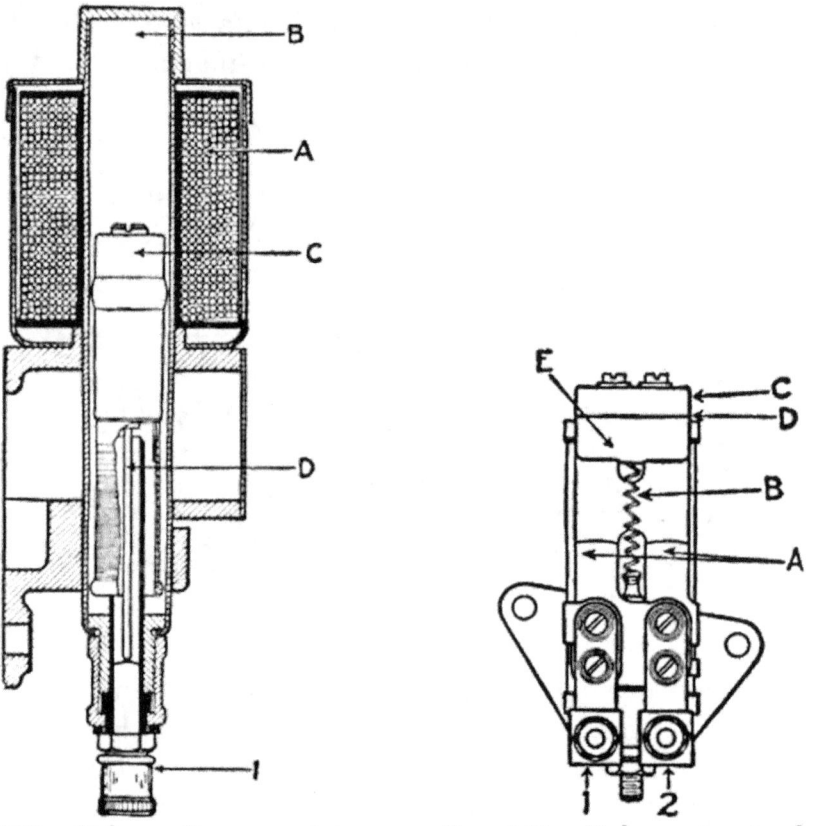

Fig. 217.—Delco voltage regulator. Fig. 218.—Delco cut-out relay.

relay is an electro-magnet with a compound winding. The voltage coil, or fine wire winding, is connected directly across the terminals of the generator. The current coil, or coarse wire winding, is in series with the circuit between the generator and the storage battery, and the circuit is opened and closed at the contacts A.

When the engine is started, the generator voltage builds up and when it reaches about 6 volts the current passing through the voltage winding produces enough magnetism to overcome the tension of the spring B, attracting the magnet armature C to core D, which closes the contacts A. These contacts close the circuit between the generator and storage battery.

The current flowing through the coarse wire winding increases the pull on the armature and gives a good contact of low resistance at the points of contact.

When the generator slows down and its voltage drops below that of the storage battery, the battery sends a reverse current through the coarse wire winding, which kills the pull on the magnet armature C. The spring B then opens the circuit between the generator and battery, and will hold it open until the generator is again started up.

139. Gray and Davis Starting and Lighting System.—The Gray and Davis starting and lighting system consists of a $6\frac{1}{2}$ volt shunt wound generator for charging the battery and furnishing current for the lights, and a series wound motor for cranking the engine.

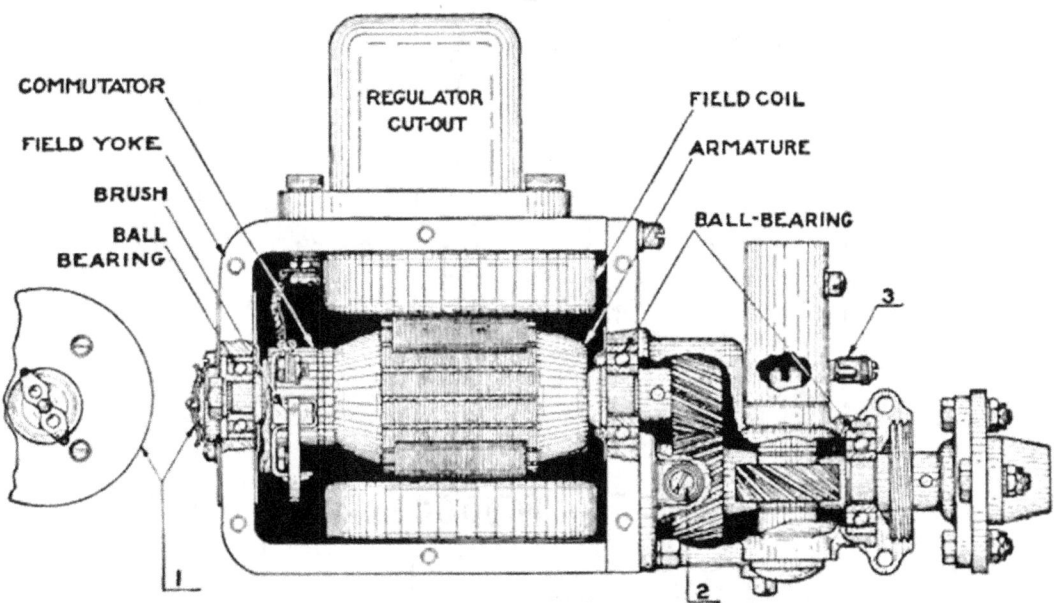

Fig. 219.—Gray and Davis generator.

The generator or dynamo is shown in Fig. 219. This generator has two shunt field windings, so arranged that the field strength or magnetism automatically increases as additional load comes on. The technical wiring diagram for the whole starting and lighting system is shown in Fig. 220.

Regulator Cut-out.—The regulator cut-out, shown in Fig. 221, performs two duties: first, to regulate the dynamo for uniform output; second, to connect the dynamo into the system only when sufficient current is generated to charge the battery and to disconnect the dynamo from the battery when the dynamo slows down so that the current is insufficient to charge the battery, and thus prevent the battery from discharging through the dynamo.

When the dynamo is at rest, the cut-out points are open and the

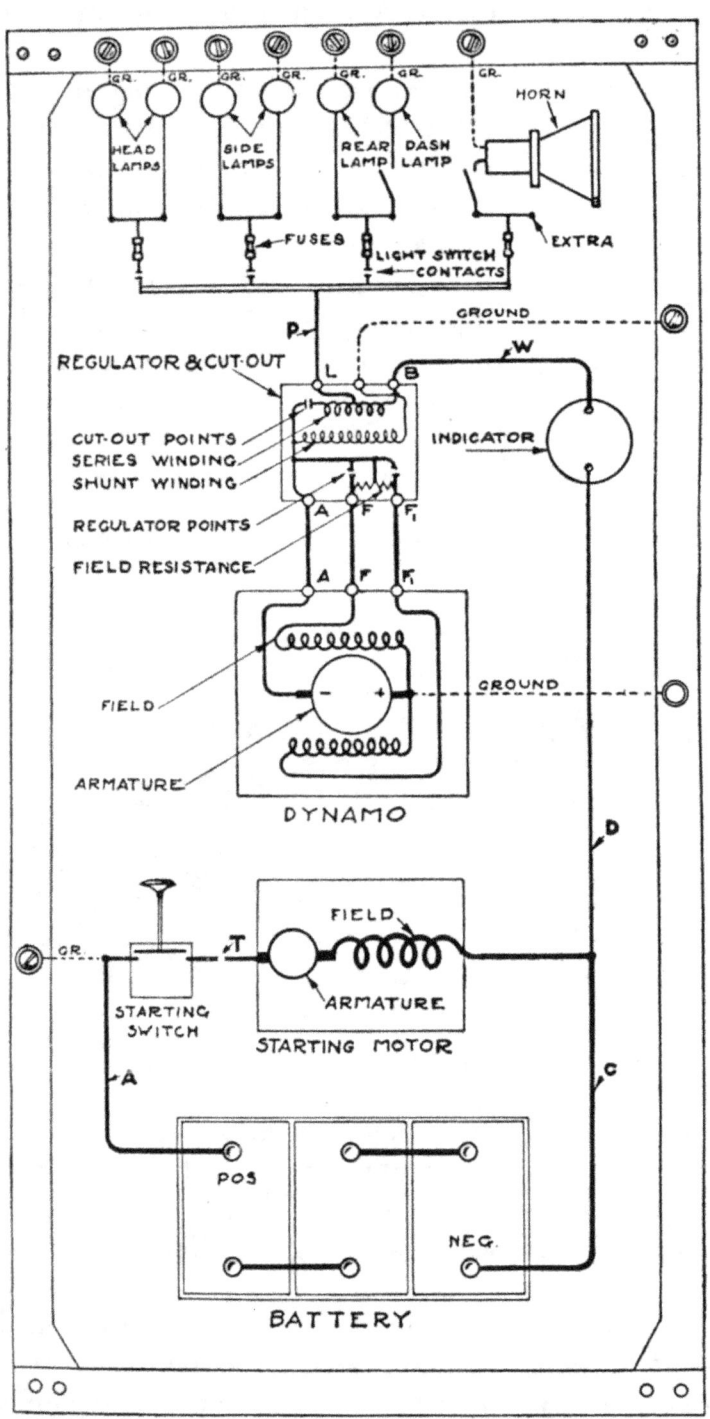

Fig. 220.—Technical wiring diagram with grounded switch—Gray and Davis starting and lighting system.

regulator points remain closed. As the dynamo first speeds up, the regulator points remain closed. Thus, the field resistance is cut out, permitting the dynamo to build up under full field strength. When the proper voltage is reached, the cut-out points close, permitting current to flow through the series winding to the system.

As the dynamo speed increases beyond that necessary for full output, the pull of the shunt winding attracts the regulator armatures. This reduces the pressure at the regulator points and inserts a resistance into the field circuit, which prevents further increase of output. The varying of the pressure at the points, which allows the resistance to be put into the circuit, is intermittent. The frequency is in proportion to the speed variation.

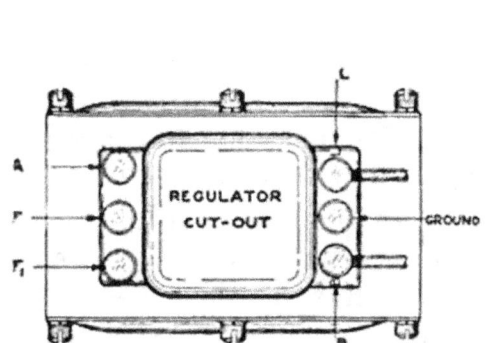

Fig. 221.—Gray and Davis regulator cut-out mounted on dynamo.

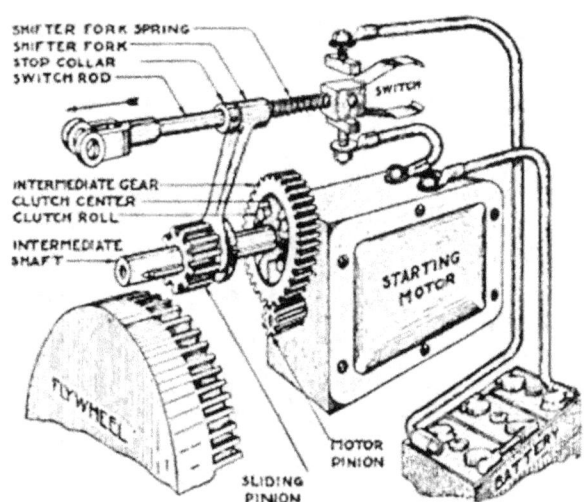

Fig. 222.—Gray and Davis starting motor and connections.

The dynamo terminals are marked B and L. B is negative $(-)$. It is the end of the regulator cut-out series winding, and connects to the battery through the indicator. L is also negative $(-)$. It is connected to the series winding at a given distance from the end and connects to the lamps through the lighting switch. The positive brush-holder of the dynamo connects or "grounds" to the dynamo frame. Therefore, the dynamo frame is positive $(+)$. Connections between the dynamo and the regulator are as follows:

The three terminals at the end of the regulator cut-out opposite the terminals marked B and L connect to the dynamo windings, as shown in the wiring diagram.

A connects to dynamo negative $(-)$ brush.

F_1 connects to the one field coil.

F connects to the other field coil.

The starting motor and its connections are shown in Fig. 222. The starting motor cranks the engine until it runs under its own power. It is the link between battery and engine. It converts electrical into mechanical energy. Electrically it is connected to the battery through heavy cables and the starting switch. Mechanically it is connected to the engine through a gear reduction having a sliding flywheel-engaging pinion and an over-running clutch.

The sliding engaging pinion and the starting switch are operated by the same operation of the starting pedal, so that electrical and mechanical connection and disconnections occur at the same time.

When the starting switch is closed, the electrical energy stored in the battery is instantly transmitted to the motor, causing the armature to rotate. This mechanical energy is transmitted through the gears and over-running clutch to the engine, causing it to rotate.

When the starting pedal is pressed to the full limit of its travel, it moves the switch rod in the direction of the arrow in Fig. 222. This moves the sliding pinion forward and closes the starting switch. If the sliding pinion is in a meshing position, it slides into mesh with the flywheel gear; but if the pinion teeth, instead of sliding between, should strike the ends of the flywheel teeth, the switch rod completes its travel, which compresses the shifter fork spring and closes the switch. When the pinion begins to turn, the compressed spring throws the sliding pinion into full engagement with the flywheel gear and permits the starting motor to crank the engine. When the engine picks up, the roll clutch prevents the engine from driving the starting motor, as the gears are in mesh until the starting pedal is released.

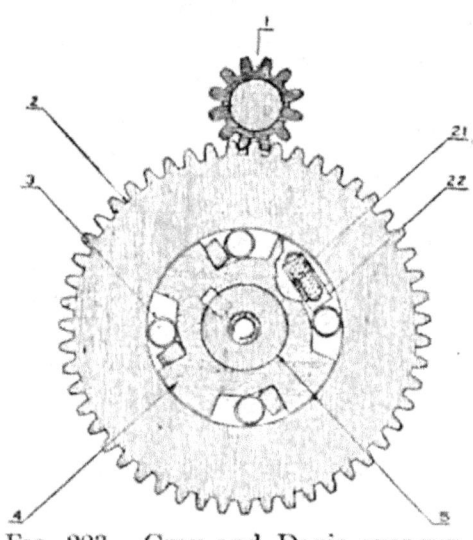

Fig. 223.—Gray and Davis over-running clutch.

Over-running Clutch.—The purpose of the over-running clutch is to permit the engine, when cranked by the starting motor, to pick up without speeding up the starting motor, which is temporarily connected to the engine while the starting pedal is pressed. This over-running clutch is merely a roller ratchet connection between one of the gears and its shaft. This is shown on Fig. 222 and is shown more in detail in Fig. 223. The gears 1 and 2 are shown in the reversed position in Fig. 223 from that which they occupy in Fig. 222.

When the starting motor pinion No. 1 of Fig. 223 is rotated in a counter-clockwise direction, the intermediate gear No. 2 rotates clockwise; the rolls No. 3 are thus rolled into the wedge angles between the curved surface of the clutch center No. 4 and the inner surface of the intermediate gear No. 2, with increased pressure until the friction is sufficient to drive intermediate shaft No. 5, which is keyed to clutch center 4.

Springs No. 21, back of the plungers No. 22, keep rolls No. 3 firmly within wedge angles so that they grip as soon as the starting motor turns. When the engine runs faster than when rotated by the starting motor, the rolls are released from the wedge angles, and the clutch center 4 can run ahead without carrying the gear 2 with it.

140. Wagner Starting and Lighting System.—The two-unit Wagner system consists of the charging generator, Fig. 224, the starting motor, Fig. 225, and the generator relay, Fig. 226. The wiring may be either the two wire or single wire system at the option of the manufacturer.

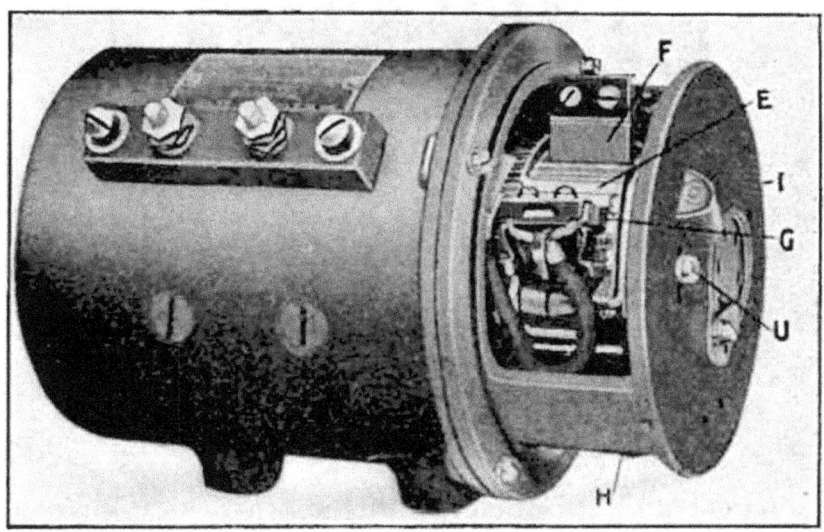

Fig. 224.—Wagner generator.

The method of connecting the generator to the engine may be by a silent chain or by spur or spiral gears. The starter motor may be connected to the engine shaft by chain and over-running clutch, or by pinion meshing with the flywheel and operated by the Eclipse Bendix system, similar to the Westinghouse clutch shown in Fig. 229. The starting motor turns the engine over at about 100 r.p.m., which is fast enough to start on most magnetos.

In Fig. 224, E is the commutator and F, G, H, and I are the brushes. The brushes H and I collect the current from the commutator and furnish this current for charging the battery through the relay. The brushes F and G collect from the commutator the current used for exciting the fields.

The function of the relay, Fig. 226, is to connect the battery to the generator when the voltage of the generator is slightly above the voltage of the battery. It also disconnects the generator from the battery when the voltage of the generator falls below the voltage of the battery. This relay consists of two magnet coils, L and M, wound on an iron core N. This iron core attracts and repels an iron lever O. At the end of this lever O are two main contact points P and Q at which the contact between the generator and battery is made and broken. There are also supplied two auxiliary contact points R and S which are for the purpose of minimizing the sparking at the main contact points P and Q. The coil M, called the shunt coil, is connected directly across the two brushes H and I, Fig. 224, and therefore the full generator voltage is impressed across the ends of this coil. The coil L, called the *series coil*, is connected in series with the

Fig. 225.—Wagner starting motor.

battery and generator and therefore this coil carries the charging current when the battery is being charged.

The action of the relay is as follows: When the engine is started, the generator is driven by the engine, and it, therefore, increases and decreases in speed with the engine. When the engine is speeded up, the generator follows with corresponding increase in speed and the voltage of the generator rises as the speed increases. As soon as the generator voltage gets to a point above the voltage of the battery, which is approximately 6 volts, the coil M, Fig. 226, pulls the iron lever O toward the magnet core, thereby closing the contact at the points P-Q and R-S. As soon as this contact is made, the generator is connected to the battery, and a charging current will flow from the generator to the battery through the series coil L, which is in series with the generator and battery. The generator continues to charge as long as these contact points P-Q and R-S remain together, but when the engine speed is decreased, so that the generator voltage falls below the battery voltage, the battery will discharge through the generator and therefore through the coil L. This discharge current, being in the *opposite direction* from the charging current will neutralize the effect of coil M and allow the spring T to pull the lever O away from the magnet core, thereby opening the contact at the points P-Q and R-S. As soon as these contacts open, the battery is "off charge." The engine speed at which this relay closes corresponds to a car speed of 7 to 10 miles per hour.

Studebaker automobiles use the Wagner system and are equipped with an instrument called a *Battery Indicator* or *Tell-tale*. This instrument is installed on the dashboard of the car and is connected in the battery circuit. The tell-tale gives indication of battery current, showing

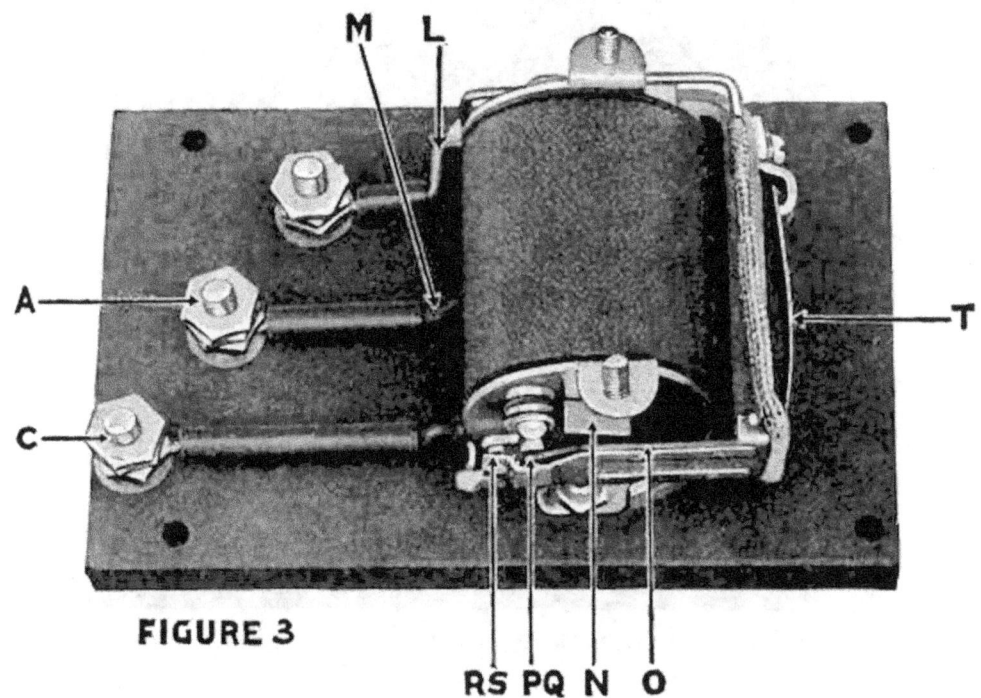

Fig. 226.—Wagner relay.

off when no current is being taken from, or being put into, the battery; *discharge* when current is being taken out of the battery by lights, ignition, or horn; and showing *charge* when the generator is charging the battery.

141. The Westinghouse Single-unit System.—The Westinghouse electric starter-lighter equipment consists of a motor-generator. In the motor-generator the functions of both starting and lighting are combined in one machine. A 12-volt system is used. The motor-generator is permanently geared or chain-connected to the engine. When the circuit is closed by the starting switch, the motor windings take current from the battery and drive the engine until firing takes place. The motor-generator is then driven by the engine, and, as speed increases, it soon generates battery voltage. At all higher speeds it charges the battery and furnishes the current for the lights.

There is an emergency feature on the Hupmobile that prevents stalling of the engine. At low speeds the motor-generator acts as a motor and assists the engine, causing an immediate restart in case of stalling.

It should be remembered that at speeds of less than 9 miles per hour, *with engine on high gear* the motor-generator acts as a motor, assists in propelling the car, and therefore takes current from the battery; and if such running is indulged in to any extent the battery will become exhausted. Also, allowing the engine to idle at low speeds will discharge the battery. A little care in avoiding low speeds and engine idling will

Fig. 227.—Westinghouse Ford outfit.

prevent this. Figure 227 shows the Westinghouse single-unit system for Ford cars, while the wiring diagram is given in Fig. 228.

142. Westinghouse Two-unit System.—The starting motor for the Westinghouse two-unit system is shown in detail in Fig. 229. It may be equipped with either a non-automatic or an automatic pinion-shift, flywheel drive.

Figure 230 shows the mechanical and electrical connections of motor and switch for non-automatic pinion-shift, flywheel drive. At A is shown the "off" position of the shift pinion and switch contactor. Pressure on the starting lever moves the shift rod first to the position shown in B, closing the motor circuit at P and P' through the resistance R; this starts the motor at low speed. Further motion of the shift rod to position C opens the electric circuit but the motor and pinion continue to turn, owing to their momentum. When position C is reached, the pinion is still turning slowly, so that it can not fail to mesh with the gear, but as power is turned off the motor, there is no difficulty in sliding the teeth into full engagement. As soon as the teeth do engage, further foot pressure on the starting lever shifts the rod to position shown in D, closing the electric circuit at Q after the pinion and gear have meshed a sufficient distance to present a good bearing length on the teeth; this

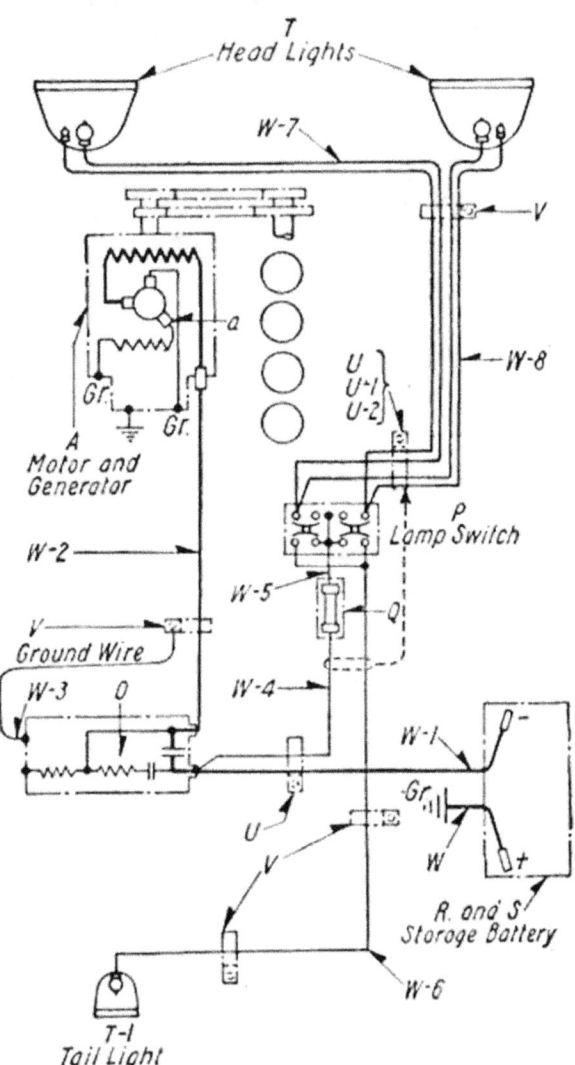

Fig. 228.—Wiring diagram for Westinghouse Ford outfit.

connects the motor directly to the storage battery so that full power is impressed, and it turns the engine over until the starting lever is released or the engine picks up on its own power. There is an over-running clutch between the flywheel pinion and the motor, so that, if the pedal is not promptly released when the engine picks up, the motor is not driven by the engine.

202 THE GASOLINE AUTOMOBILE

Fig. 229.—Westinghouse starting motor disassembled.

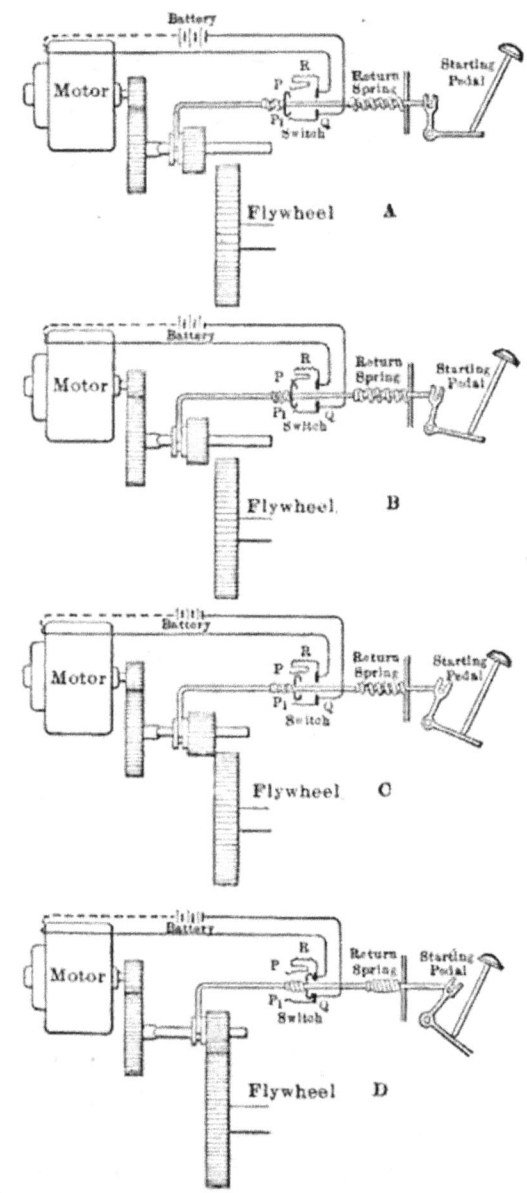

Fig. 230.—Connections of Westinghouse starting motor with non-automatic pinion shift.

STARTING AND LIGHTING SYSTEMS

In the Eclipse-Bendix pinion-shift, as shown in Fig. 231, the starter motor is fitted with a special threaded shaft which automatically shifts the pinion into mesh with the flywheel when the starting switch is closed. When the switch is closed, the full battery voltage is impressed on the motor, and it starts immediately. The pinion, when the motor is at rest, is within the screwshaft housing and entirely away from the flywheel gear. The threaded shaft is connected to the reduction gear shaft by a spring which thus forms a flexible coupling. As the load is not large enough to compress the spring when the motor starts, the threaded shaft is immediately revolved by the spring in released position. The pinion moves out on its shaft by virtue of the revolving threads, until it reaches the flywheel. If the teeth of the pinion and flywheel meet instead of

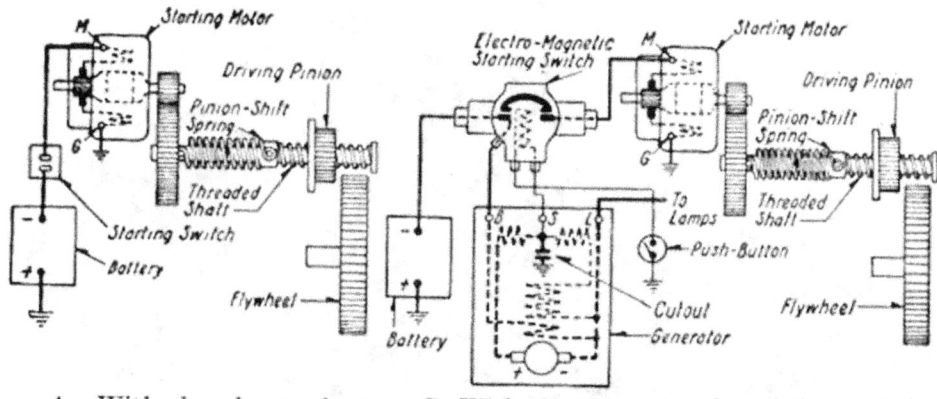

A, With hand or foot operated starting switch. B, With electro-magnetic starting switch controlled by push button.

Fig. 231.—Connections of Westinghouse starting motor with Eclipse-Bendix pinion shift.

meshing, the spring allows the pinion to revolve until it meshes with the flywheel. When the pinion is fully meshed into the flywheel teeth, the spring compresses, and the pinion is then revolved by the motor as through a continuous shaft, turning the engine over. When the engine fires and the peripheral speed of the flywheel continuously exceeds that of the driving pinion, it forces the latter out of mesh, and it is returned to its original position in the screwshaft housing.

The Westinghouse lighting and starting generator, as shown in detail in Fig. 232, is operated by belt, chain, or gear drive from the engine and furnishes current to the storage battery and lights. While the engine is stopped or running at very low speed, the lights are supplied entirely by the battery. A magnetic switch in the generator automatically connects the generator to the lighting system and battery when the engine is running at approximately 8 miles per hour car speed on direct drive. When running on the gears, the switch closes at a much lower car speed. If no lights are then in use, the battery begins to be charged when this switch makes the electrical connection. If the lights are burning, the generator furnishes part of the current to them; as the speed increases, the proportion of current supplied by the generator increases, until at high speed the generator supplies all of the current to the lights and in addition charges the battery. The amount of current the generator furnishes to the battery depends upon the number of lamps burning and upon the speed of the engine.

143. The U. S. L. Electric Starting and Lighting System.—This is a unique system in which a single unit motor-generator is connected directly to the engine shaft, taking the place of the flywheel.

The motor-generator consists of a stationary housing, a set of fields complete with poles and coils, an aluminum case, and a dust ring, as

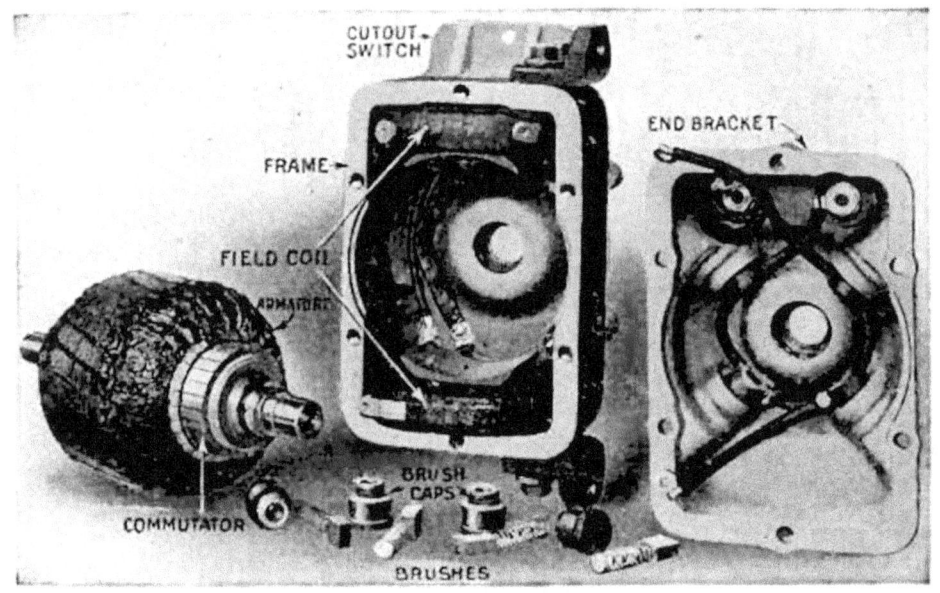

Fig. 232.—Westinghouse starting and lighting generator disassembled.

shown in detail in Fig. 233. The armature replaces the flywheel of the engine, being attached to the crank shaft in its stead as shown in Fig. 234. When the starting button is pressed down, the current from the battery starts the electric motor. This revolves the crank shaft of the engine. With the switch of the ignition coil in battery position, the explosions will commence. The starting button should be quickly released, thus automatically changing the electric motor into an electric generator. As the speed of the engine increases, the generator gradually commences charging the battery, restoring the current discharged during the starting operation.

Regulator.—The regulator is located on the dash under the cowl and instrument board. It performs four principal functions: 1, closes the switch when the generator voltage is sufficient to charge the battery; 2, opens the switch when the generator voltage is insufficient and the current reverses; 3, regulates the maximum charge to the battery; and 4, controls the generator voltage on open battery circuit.

An indicating arrow is visible through the window in the regulator cover when the switch is closed and the storage battery is being charged. It disappears when the contact is broken. The switch should close when a car speed of 10 to 12 miles per hour is attained, and open when the speed falls below about the same rate, or when the motor stops altogether.

The regulator consists of a magnet coil which pulls the switch lever into contact when the proper car speed is attained. It also acts on a carbon pile lever and controls the field current by increasing or decreasing the resistance through the carbon discs at the top of the regulator.

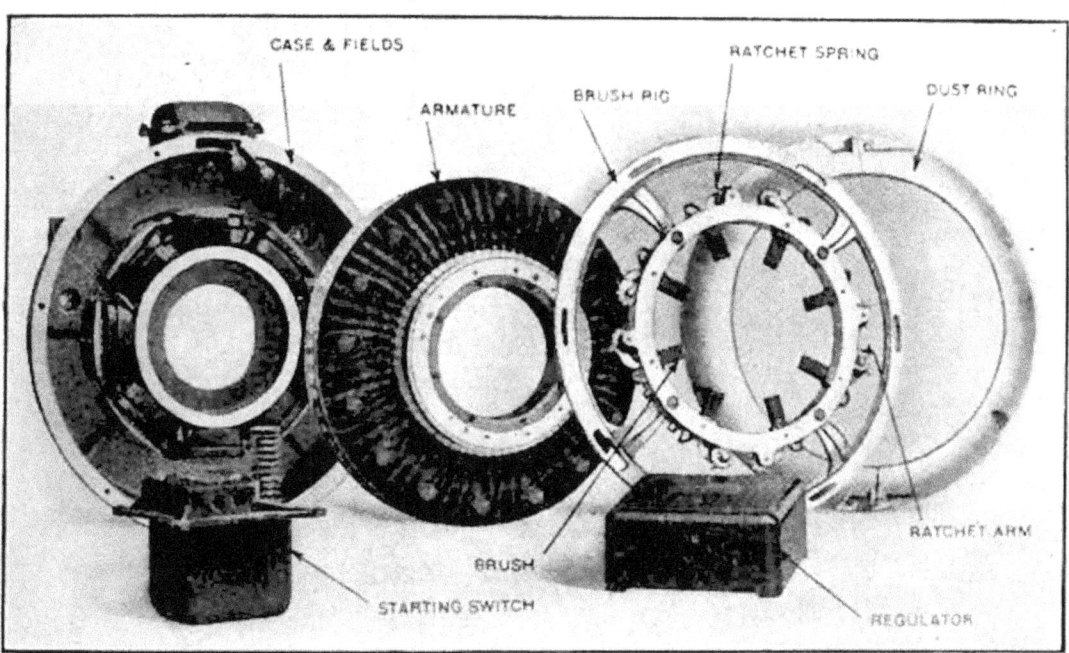

Fig. 233.—Details of U. S. L. system.

If the engine does not turn over when you first press on the button, immediately let up the button and try again several times quickly. Do not hold your foot on it long, as this will needlessly drain the current from the battery.

If the motor fails to respond when the starter button is pressed several times quickly, the battery is too low. In such cases, do not continue to hold the starting button down; release it, and crank the motor by hand, running it at a charging rate of 10 to 15 amp. giving the generator an opportunity to recharge the battery. If you repeatedly press the starting button without running the engine, it will only be a question of time before the battery will be exhausted.

144. Jesco Single-unit Electric Starter and Lighter.—The complete system consists of a starter-generator, with controller and starting switch mounted thereon, in connection with a 6 volt, 100 amp.-hour storage battery, switch, and wiring for lights. The starter-generator is connected either by coupling, by silent chain, or by gears to the crank shaft of the engine, at a ratio of either one to one, or two to one.

The electric machine performs as a series motor at time of starting and as a shunt generator for storing current in the battery and supplying the lights. As a starter, a gear reduction is automatically engaged, and

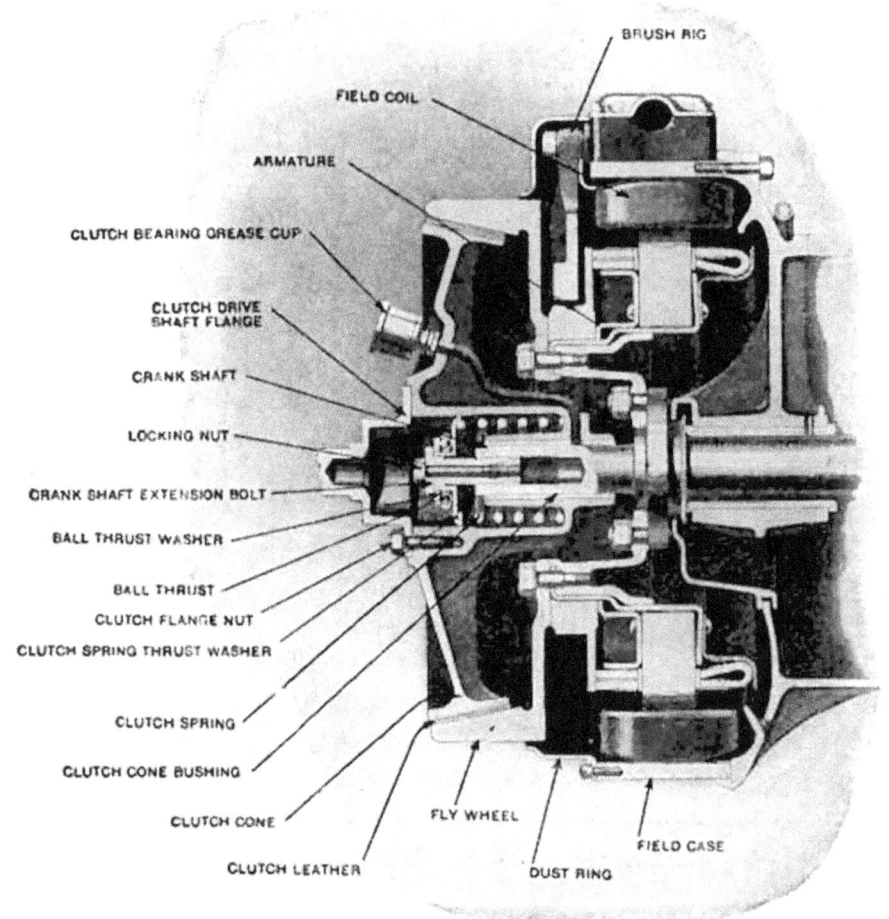

Fig. 234.—Section of U. S. L. motor-generator.

after the engine starts, this transmission locks by action of a multiple disc clutch, and no gears are in operation. This works automatically and requires little attention, outside of oiling. The electrical regulating mechanism is contained in the little box on top of the starter-generator. The regulation is taken care of by a differential shunt field in connection with an automatic regulator.

Charging begins at approximately 8 miles an hour car speed. At 15 miles an hour the maximum charging rate is reached and, by regulation, remains constant through all speeds in excess of that amount. The battery cut-out automatically disconnects the battery when the generator is not charging, preventing a back flow from battery to machine.

The wiring is extremely simple, having only two leads from starter-generator to battery, with the lighting of car and the indicating meters arranged as desired. The Jesco system as used on a Continental six engine is shown in Fig. 235.

145. Care of Starting and Lighting Apparatus.—A periodical inspection should be made of wiring, insulation and all connections. Wiring and connections should be protected against grease, oil, and mechanical injury.

FIG. 235.—Jesco starting installation.

Use the same consideration for your auto lighting system that you do for electric light in your house. Do not leave your car all night with all lights burning and expect to find a well charged battery in the morning.

Be sure that all wires are perfectly insulated and not in contact with any moving parts, as the constant rubbing will wear off the insulation and the vibration will cause the connections to become loose.

All permanent connections should be well soldered, all stray strands of wire removed and the joints properly taped in order to prevent loss of current from short circuits. If wires must be run where there is liable to be grease, oil, or water, they should be protected by conduit or other oil or waterproof material. Either oil or water will cause the insulation on the wire to be of very little value.

The generator should be inspected about every month and kept clean. The commutator may become rough and blackened and should be cleaned by holding a piece of fine sandpaper against it while rotating. Then carefully remove all metallic particles from the commutator bars that might cause a short circuit between them. A short circuit may also be caused by carbon dust from the brushes.

The brushes should always have a perfect bearing surface on the commutator. The general cause of a poor bearing is that the carbon brush sticks in the brush holder. It may be taken out and filed down so that it will slide easily in the holder.

When putting in new brushes, make sure that they fit perfectly on the commutator. It is also a good policy to use only the brushes sent out by the manufacturer of the machine.

If there is a grounded wire in the machine, or if a commutator segment becomes loose, the armature should be returned to the factory for repairs.

The carbon brushes contain sufficient lubricant for the commutator so that it is not necessary to use any oil or grease of any kind. If grease or oil should accumulate on the brushes or commutator, it should be wiped off with a dry cloth.

The starting motor is intended to perform one function only, viz., to spin the engine, and should only be used for such purpose. Any attempt to propel the car by the starting motor or indulge in the needless use of same will result in trouble. Such experiments are of no material value and it is no test of the power of the starting motor, but simply imposes an extravagant demand on the battery. If these practices are indulged in they will result in a complete discharge, which is detrimental to the life and service of the storage battery.

146. Starting Motor Troubles.—The closing of the starting switch completes the circuit between the battery and the motor, and puts the starter in operation. If the starter does not turn the engine over, release the switch at once and ascertain if all connections are tight and secure, and inspect the battery. If the starting motor turns the engine over very slowly, it is evident that the battery is weak or the engine stiff. If the starting motor is turning the engine over at a reasonable cranking speed and the engine does not fire, remember that the starting motor is performing its duty, so do not let the starting motor continue to crank the engine longer than necessary, as a needless drain is placed on the battery. If the engine does not fire, it is evident that the trouble is confined to the carburetor or ignition.

147. Generator Troubles.—A simple test to determine if the generator is properly operating is, first, to switch all the lights on with the engine idle; second, to start the engine and run it reasonably fast. If the lights brighten perceptibly after starting the engine, it proves that the generator is properly delivering current. This test must necessarily be conducted in the dark, either in the garage, or preferably, at night time. Generator troubles will be manifested by dim lights when the engine is running at a medium rate or by failure to keep the battery charged. The trouble may be caused from, first, grounds or short circuits in the field windings; second, increased resistance in circuit, caused by dirty commutator or brushes, weak brush springs or poor material in the brushes (poor material in brushes causes sparking and overheating); third, grounds in the armature, caused by defective insulation or carbon deposits on the commutator short-circuiting the copper bars; fourth, circuit breaker or regulator not properly adjusted so that the battery is not cut in at proper time. The contact points may become dirty or corroded or may be burned by an excess of current, generally from a reverse current from the battery.

148. Battery Troubles.—Battery troubles may be detected by failure to turn the motor or by the lights burning dimly when the engine is stopped. Battery troubles can be traced to improper charging; loss of electrolyte; short circuits, either external or internal; overloading, caused by using light bulbs of too large capacity; burning lights when not necessary; and propelling car with starting motor. External short circuits may be caused by broken insulation so that two bare wires come together or come into contact with the frame of the car or other conducting material, or may be caused by acid on top of battery forming circuit between terminals. Internal short-circuiting is explained in Art. 134.

If the starting motor will not crank the engine, the trouble may be looked for as follows:

1. Battery discharged.
2. Broken circuit caused by worn out or dirty brushes or weak springs, or broken connections or short circuits in any part of the wiring or switches. A dirty commutator will have the same effect as dirty brushes.

If the starting motor cranks the engine very slowly, the trouble may be caused by the battery being partly discharged or by an excess of resistance in the circuit. The increased resistance may be caused by loose connections in wiring, poor contacts in switch, dirty brushes or commutator, brushes made from unsuitable material or not held firmly on the commutator.

149. Winter Care of Batteries.—If the car is not to be used for some time, as in the winter, the batteries should be inspected before the car is used for the last time. Water should be added to the cells, if necessary, so that it will thoroughly mix with the electrolyte when the car is driven. When the car is laid up, the specific gravity of the electrolyte should register from 1.27 to 1.29. In this condition there will be no danger of freezing in any climate. The battery should be charged every two months during the "out of season" period, either by running the engine, or by charging from an outside source. If either of the above methods is impossible, and there is no garage convenient that is equipped for charging batteries, the battery may be allowed to stand without charging during the winter, providing it is fully charged at the time the car is laid up. Much better results, however, and longer life of the battery will be obtained by giving the periodic charges. The wires of the battery should be disconnected during the "out of season" period in order to prevent any slight leaks that might occur in the wiring.

150. "Don'ts" on Starting Equipment.—Don't disconnect the battery and start the engine up with any of the lamps in circuit. This is very important as the battery acts as a voltage regulator and, if not connected, the lamps or fuses in the circuit connected will be blown out immediately, due to heavy rise in voltage from the generator.

Don't attempt to work around the lighting system without disconnecting the battery ground and winding it with tape. It is a very easy matter to touch a screw-driver or a pair of pliers from a live wire to the frame or to the pipes or engine, thereby causing short circuit and blowing out a fuse. When the work is finished, replace the ground wire before starting the engine.

Don't try to repair or readjust any of the instruments supplied. Leave this to the manufacturers whose experience in this field will insure handling the job in a better manner than you can.

Don't leave the starter button in the socket while the motor is running.

Don't stamp on the starter button, but press it down deliberately and firmly.

Don't fail to go over the wiring occasionally and see that all binding posts are tight and free from corrosion.

Don't fail to remember that the mechanism is an electrical starter and not a motor for vehicle propulsion.

Don't expect the starter to spin the motor at a maximum cranking speed if the battery voltage is run down. Endeavor to run the car with fewer lights for a while and allow the voltage to pick up.

Don't abuse the electric starter. The mechanism is strong and durable and guaranteed for the purpose intended, but is not guaranteed against rough treatment or inexcusable abuse.

Don't fail to inspect all terminals occasionally and see that the tape which protects these terminals from short-circuiting is in good shape. In case this has become unwrapped, it is advisable to replace immediately with fresh insulating tape of good quality.

Don't try to hook up additional electrical equipment without carefully going over the wiring diagram to find the proper place for such a connection.

Don't fail to see that the ground wire from the battery has a good contact between the terminal and frame.

Don't fail to carry extra fuses and lamp bulbs.

CHAPTER IX

AUTOMOBILE TROUBLES AND REMEDIES

151. Classification of Troubles.—The manufacturers of automobiles are constantly striving to simplify the design and construction of all parts in order to reduce the number of troubles which are a constant source of worry to the automobile owner and driver. They have been

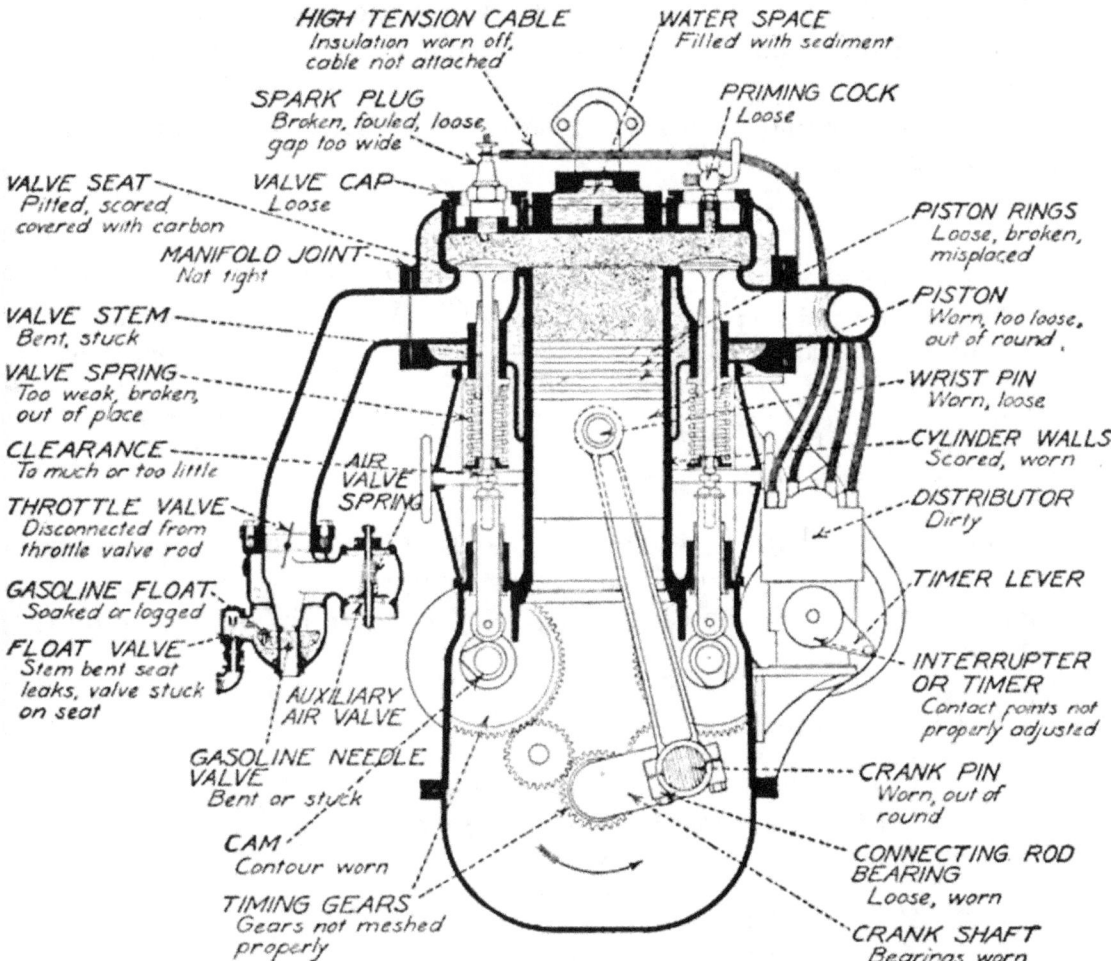

FIG. 236.—Chart showing location of common mechanical troubles of engines.

quite successful in reducing troubles to a minimum; as a matter of fact, the possible troubles on the modern car are now few in number compared to those of not a great many years ago. The troubles now commonly experienced are those inherent in every man-made machine which is subject to the wear and tear of everyday use.

It is obviously impossible in many cases to give a direct statement of a cure for all of the various symptoms which are likely at some time or other to confront the motorist, as some symptoms may be due to any one or more of several different causes. All that can be done is to offer a few general suggestions which will assist him to diagnose his own specific troubles and apply the proper remedy.

The automobile is a fine piece of machinery and the service from it will depend upon the care and attention given to it. Many of the troubles on the modern automobile are due to uncalled for adjustments and investigations by the motorist. Although good care and attention must be given in order to get efficient service, it is good policy to leave well enough alone and not do any unnecessary tampering, nor try to improve upon the operation or construction as planned by the manufacturer.

The more common motor car troubles can be divided into the following general headings:

I.	II.	III.
Power plant troubles	Transmission troubles	Chassis troubles
(a) Mechanical parts of engine.	(a) Clutch.	(a) Wheel hubs.
(b) Carburetting and gasoline system.	(b) Change gears.	(b) Steering gear.
	(c) Differential.	(c) Brakes.
(c) Ignition.	(d) Rear axle.	(d) Springs.
(d) Lubricating and cooling.		(e) Tires.
(e) Starting and lighting.		

152. Power Plant Troubles.—Any derangement in the power plant will show itself by one of the following symptoms. Under each symptom is given the common causes with a reference to the discussion on the subject.

(1) *Engine Fails to Start.*
    (a) Poor compression. See Art. 153.
    (b) Engine cylinder flooded. See Art. 154(e).
    (c) Carburetor adjustment not right. See Art. 154.
    (d) Water in gasoline. See Art. 154(j).
    (e) Carburetor frozen. See Art. 154(g).
    (f) Out of gasoline. See Art. 154(i).
    (g) Engine too cold. See Art. 154(f).
    (h) Ignition switch off.
    (i) Foul or broken plugs. See Art. 155(b).
    (j) Weak batteries or magneto. See Art. 155(e, f, and g).
    (k) Vibrators not properly adjusted. See Art. 155(h).
    (l) Wiring system out of order. See Art. 155(d, j, and k.)

(2) *Engine Misses at Low Speeds.*
    (a) Poor compression. See Art. 153.
    (b) Mixture too lean or too rich. See Art. 154(a and b).

AUTOMOBILE TROUBLES AND REMEDIES

    (c) Spark plug gap too wide. See Art. 155(b).
    (d) Spark plug cable not connected or short-circuited. See Art. 155(d).
    (e) Dirty interrupter. See Art. 155(k).
    (f) Dirty or defective spark plug. See Art. 155(b).
    (g) Vibrator not properly adjusted. See Art. 155(h).

(3) *Engine Misses at High Speeds Only.*
    (a) Carburetor not set for this speed. See Art. 154(a and b).
    (b) Bad spark plug. See Art. 155(b).
    (c) Weak valve spring. See Art. 154(b).
    (d) Timer contact imperfect. See Art. 155(k).
    (e) Vibrator points dirty or burned. See Art. 155(h).

(4) *Engine Misses at All Speeds.*
    (a) Carburetor not properly adjusted. See Art. 154(a and b).
    (b) Dirty or broken plug. See Art. 155(b).
    (c) Spark plug gap not right. See Art. 155(b).
    (d) Poor compression. See Art. 153.
    (e) Loose or broken terminals. See Art. 155(d).
    (f) Weak batteries or magneto. See Art. 155(e, f, and g).
    (g) Defective wiring. See Art. 153(d).
    (h) Coil not properly adjusted. See Art. 155(h).
    (i) Gasoline feed stopped up. See Art. 154(b and h).
    (j) Needle valve bent or stuck. See Art. 154(b and h).
    (k) Water in gasoline. See Art. 154(j).
    (l) Poor circulation. See Art. 156(b).
    (m) Excessive lubrication. See Art. 156(a).

(5) *Engine Overheats.*
    (a) Lack of proper circulation. See Art. 156(a).
    (b) Lack of proper lubrication. See Art. 156(a).
    (c) Slipping fan belt or bent fan blades. See Art. 156(b).
    (d) Too rich a mixture. See Art. 154(a).
    (e) A weak mixture. See Art. 154(b).
    (f) Running with spark retarded. See Art. 155(l).
    (g) Carbon deposit in cylinders. See Art. 153(f) and 155(m).

(6) *Engine Stops.*
    (a) Gasoline tank empty. See Art. 154(i).
    (b) Water in gasoline. See Art. 154(j).
    (c) Carburetor flooded. See Art. 154(d).
    (d) Lack of pressure on gasoline tank. See Art. 154(i).
    (e) Overheating due to poor circulation or lack of lubrication. See Art. 156(a and b).
    (f) Short-circuiting of wires or terminals. See Art. 155(d and j).
    (g) Disconnected or broken wires. See Art. 155(d).
    (h) Wet batteries or magneto. See Art. 155(d and e).

(7) *Engine Knocks.*
    (a) Carbon deposits in cylinder and on piston heads. See Art. 153(f) and 155(m).
    (b) Spark too far advanced. See Art. 155(l).
    (c) Running motor slow when pulling heavy load on direct drive. See Art. 155(l).
    (d) Faulty lubrication. See Art. 156(a).
    (e) Engine overheated. See Art. 155(m).
    (f) Loose connecting rod bearings. See Art. 153(g).

(g) Loose piston. See Art. 153(e).
(h) Loose crank shaft bearing. See Art. 153(g).

(8) *Engine Will Not Stop.*
   (a) Short circuit in switch.
   (b) Magneto ground may be disconnected.
   (c) Overheating and carbon deposits. See Art. 155(m)

(9) *Lack of Power.*
   (a) Poor compression. See Art. 153.
   (b) Too weak or too rich a mixture. See Art. 154(a and b).
   (c) Weak spark. See Art. 155(e, f, g, and h).
   (d) Lack of lubrication. See Art. 156(a).
   (e) Lack of cooling water. See Art. 155(b).
   (f) Lack of gasoline. See Art. 154(h and i).
   (g) Dragging brakes. See Art. 159(c).
   (h) Slipping clutch. See Art. 158(a).
   (i) Flat tires.
   (j) Choked muffler causing back pressure.

(10) *Back-firing Through Carburetor.*
   (a) Improper needle valve adjustment. See Art. 154(b).
   (b) Dirt in gasoline passage or nozzle. See Art. 154(b and h).
   (c) Inlet valves holding open. See Art. 154(b).
   (d) Excessive temperature of the hot water jacket of the carburetor, especially in hot weather. This can be remedied by shutting off the water from the carburetor jacket and cutting off the hot air supply.
   (e) Spark retarded too far. See Art. 154(b) and 155(m).

(11) *Firing in Muffler.*
   (a) Weak mixture, slow burning exhaust, igniting unburned charge from previous "miss." See Art. 154 (b).
   (b) Valves out of time.
   (c) Too rich a gasoline mixture. See Art. 154(a).
   (d) Occasional missing of a cylinder.

(12) *Starter Will Not Operate.*
   See starter troubles, Chap. VIII.

153. Mechanical Troubles in Engine.—(a) *Poor Compression.*—Poor compression is one of the common causes for lack of power. Unless the compression pressure is high enough, the explosion will be lacking in force and the engine will be weak. The engine can be turned by hand, with the ignition off, throttle open, and the compression noted in each cylinder, or a more accurate way is to remove the spark plug and screw in a small pressure gauge, which should show from 60 to 80 lb. at the end of the compression stroke, depending on the make of engine. Loss of compression is commonly due to leaky or improperly seated valves, or to leaky joints. Leaky thread joints, valve caps, and cracks in cylinder are common causes for loss of compression. These can be detected by a hissing sound or, if the suspected leak is covered with gasoline or oil, the leak will show itself by bubbling through the oil. If the trouble can not be located in this manner attention should be given to the valves.

As a rule, the intake valve requires less attention than the exhaust valve, because the former comes into contact with the cool fresh fuel charges, whereas the latter is apt to become fouled and burnt by the hot and dirty exhaust gases. A frequent cause of leaky valves is carbon deposit on the valve seats. These deposits prevent the proper seating of the valve. The remedy is to clean and grind them.

b. Grinding Valves.—There are several good grinding compounds on the market. It is advisable to use a coarse grade in the first operation and then to finish off with a finer one to give a polished surface. A very good homemade mixture is obtained by making a thin paste of a couple of tablespoonfuls of kerosene, a few drops of oil, and enough fine flour emery to thicken to the consistency of paste.

The valve spring must be removed so that the valve may be lifted and turned. A moderate coating of the paste is applied to the bevel face of the valve. Next rotate valve back and forth until the entire bearing surface is polished bright and smooth the full width of the face. The valve should never be turned the whole way round. Rotate it back and forth a quarter turn at most under light pressure, lifting it up frequently and turning it halfway round before seating it again. This method distributes the friction evenly and eliminates the possibility of the emery scoring the bearings. If no valve grinding tool is available, the use of a carpenter's brace or bit stock is recommended, as a much smoother movement is thus obtained than by using a screw-driver. This method, recommended by the Overland Company, is shown in Fig. 237.

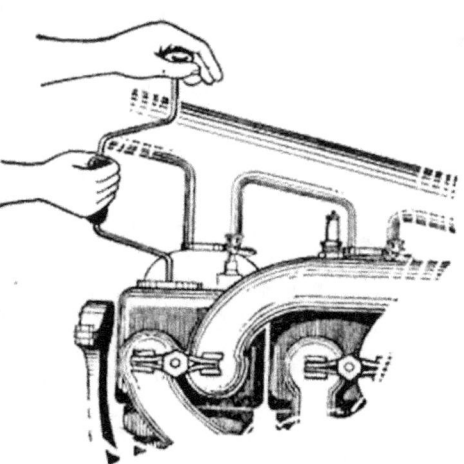

Fig. 237.—Valve grinding.

After grinding to a good clean seat entirely free from spots or pits, wash the valve, valve seat, and guide thoroughly in gasoline. If the stem is rough or gummy, smooth it up with emery cloth but clean it afterward before replacing it in the guide. To test the effectiveness of your work, mark the valve seat in several places with a lead pencil and turn the valve around a few times. If the marks are entirely rubbed off, the work may be considered well done.

(c) *Valve Adjustments.*—Poor adjustments of the valve operating mechanism may cause poor compression, even if the valve seats have been properly ground in. The valve spring may be broken or too weak to close the valve on its seat in the proper time. Sticking of the valves when open may also be the cause of low compression.

The clearance between the valve stem and push rod may be the cause of considerable trouble. This clearance is usually about the thickness of a thin visiting card, the exact amount being somewhat different for different cars, but never over $\frac{1}{32}$ in.

If this clearance for the intake valve is too great, the lift is reduced, thus preventing the proper charge from getting into the cylinder. If the exhaust valve lift is reduced in the same way, it will be more difficult for the exhaust gases to escape. Too much clearance also changes the time of

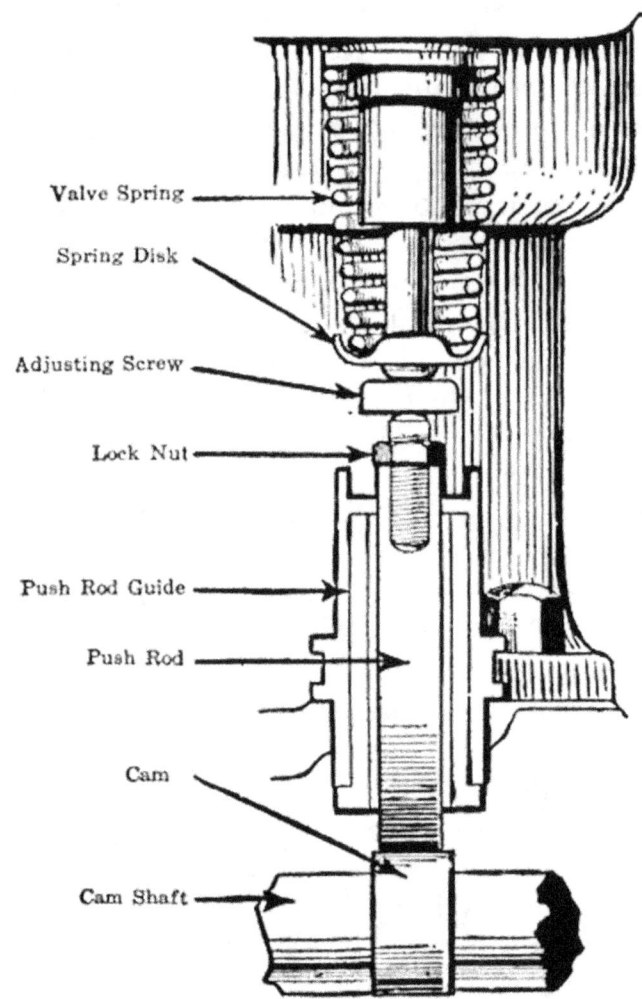

Fig. 238.—Adjustment of push rod clearance.

valve opening and closing, causing the valves to open late and close early. If, on the other hand, this clearance is too small or entirely absent, the valve will open early and close late, or will not close on its seat at all.

As the valve seats are lowered by continual grinding, the clearance is gradually changed. For the proper operation of the valves, careful attention should be given to this clearance space. Figure 238 illustrates the clearance adjustment on the Overland car.

A weak spring on the exhaust valve may have a marked effect on the operation of the engine. The exhaust valve then opens on the suction stroke and burnt gases are again drawn into the cylinder.

(d) *Valve Timing.*—It is essential that the valves be properly timed or set, in order to have the engine operated properly. The valves are set at the factory and the necessity for adjusting the timing comes as the result of wear on the valve seats, stems, rods, cams, half-time gears, or by improper replacement of any of these parts. If the cam shaft has been removed, care must be taken to get the gears properly meshed when replacing it. The gears are marked so that replacement is not difficult. The proper method of replacing the gears on the Ford engine is shown in Fig. 239. It will be noticed that there is a prick-punch mark on one tooth of the pinion and a corresponding mark on the large gear. Before taking a cam shaft out, an examination should be made and if the gears are not so marked it should be done before they are disturbed.

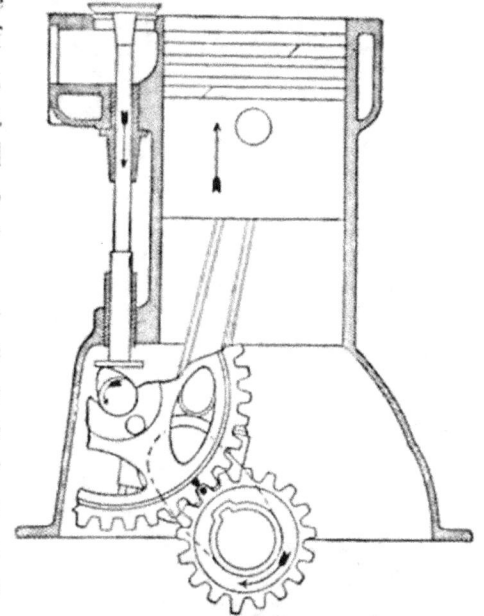

Fig. 239.—Ford cam shaft setting, showing marked tooth and space on timing gears.

If the clearances are properly adjusted for the push-rods and valve stems and if the timing gears are properly meshed, the valves should be correctly timed, making allowance for wear on the cam faces. Most engines have the positions at which the valves start to open and close marked on the circumference of the flywheel. These points should be opposite the pointer, usually at the top of the case, when the valves start to open and close. This time can be determined by the use of a thin sheet of tissue paper. By placing a piece of the paper in the clearance space between the push-rod and valve stem, one can tell when the valve opens or closes.

Valve setting is an adjustment that should be made by an experienced mechanic or one thoroughly familiar with the principles of the four-stroke engine. The different makers have found by trial the settings that will give the best results with their engines and cars. These settings differ somewhat according to different conditions. If they are not marked on the flywheel, they should be obtained from the manufacturer.

Figure 240 shows the approximate crank and piston positions for the valve events. The inlet may be opened by the different makers anywhere from top center to 20° of flywheel motion after center. The inlet closes from 25° to 50° past lower center. The exhaust opens 35° to 60° before lower center and closes from top center to 15° past center.

(e) *Loose Piston or Scored Cylinder Walls.*—A loose piston or scored cylinder walls will cause a marked loss of compression. If the piston is

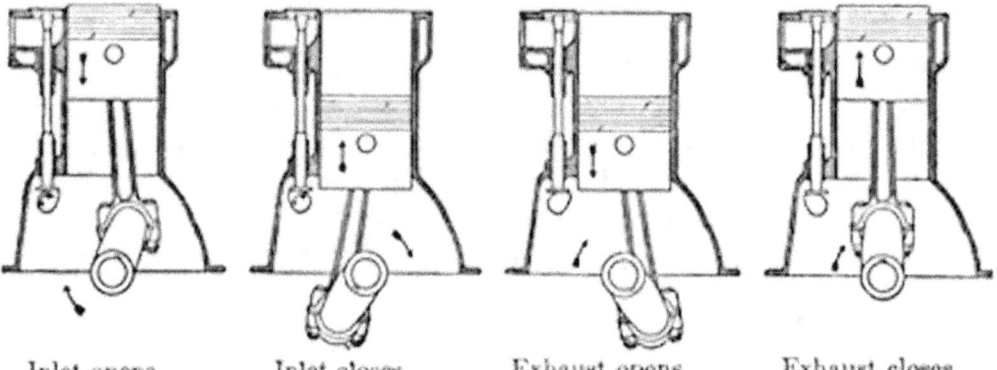

Inlet opens.   Inlet closes.   Exhaust opens.   Exhaust closes.

Fig. 240.—Valve setting diagram.

not too loose, slightly larger rings may be put on. Sometimes the blowing can be remedied by using a heavier cylinder oil. This will to some extent remedy the trouble caused by scored cylinder walls, although if too badly cut, they must be rebored and new pistons or rings fitted in. Again, this is the work of an experienced mechanic.

(f) *Carbon Deposits in Cylinder.*—After the engine has been run for some time, carbon deposits are liable to collect in the cylinder, and on the pistons, especially if too much lubricating oil or gasoline has been used. The carbon deposit resulting from too much lubricating oil is a sticky substance, while that from too much gasoline is hard, dry, and brittle. These deposits, if allowed to collect, become hot from the heat of explosions, and cause preignition of the fresh charge of gas.

The best methods of removing carbon deposit are to scrape it out or to burn it out by means of an oxygen flame. The latter method is quicker and by far the most convenient. The following method is recommended by the Overland Company for the removal of carbon by scraping:

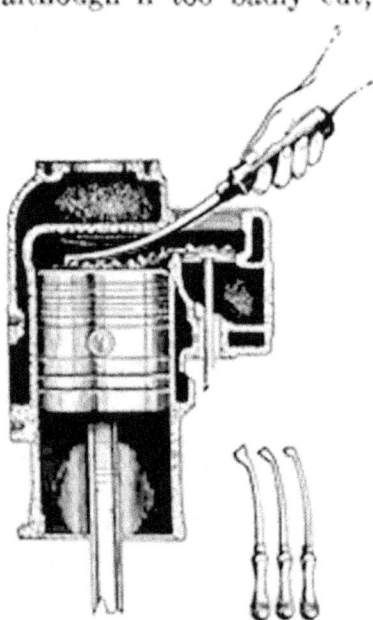

Fig. 241.—Scraping the cylinders.

To scrape the cylinders, remove both inlet and exhaust valve caps. Fig. 241, and turn the motor over until the pistons of two cylinders are at their top centers. The scraping off of the deposit is done by means of tools of different shapes, the tools being bent so as to reach the piston head and the sides and tops of the cylinders. Scrape all removed carbon over to the exhaust valve and, when through, turn the motor until the exhaust valve lifts, when the carbon may be scraped past the valve and into the exhaust passage, whence it will be blown out. For a good job, brush the surfaces clean and make sure that no carbon becomes lodged between the exhaust valve and its seat. Finally wash with kerosene.

In replacing the cylinder plugs over the valves, put graphite grease around the threads; this will make a compression-tight joint and also make it easier to remove the plugs the next time. Likewise, be sure to replace the copper gaskets under the plugs.

It is an excellent plan to attend to removing the carbon and to grinding the valves together at the same time.

Kerosene is also used for the removal of carbon from the cylinders. Pour two or three tablespoonfuls of kerosene through the priming cocks while the engine is warm. It has a strong solvent action on any gummy binding material in the carbon and can be spread over the entire cylinder by cranking the engine a few times around. Some motorists inject the kerosene through the air valve of the carburetor just before the engine is stopped preparatory to putting it away for the night. Kerosene will not remove a hard carbon deposit but it will prevent it from forming if used regularly about once a week.

Running the engine on alcohol for a few minutes is another device that is sometimes used for burning out carbon deposits.

(g) *Bearing Troubles.*—The common bearing troubles are those caused by the bearings becoming worn and loose, with a consequent knocking. Faulty lubrication, clogged oil pipes and oil holes, and dirty oil are the main causes of warm bearings. The bearings which are most liable to give trouble are the wrist pin bearings, the connecting rod bearings, and the main crank bearings. After a bearing has been excessively hot, it should be refitted by a mechanic. A loose bearing can be tightened on the pin by removing the liners or shims, or by being refitted.

154. Carburetion Troubles.—Improper mixture is the common source of carburetor trouble. The mixture is either too rich, that is, too much gasoline in proportion to the air, or too weak, that is, too much air in proportion to the gasoline.

(a) *Mixture too Rich.*—A rich mixture shows itself by black smoke coming from the muffler, and by overheating and missing of the engine. Not only is fuel wasted, but the cylinders become fouled and carbonized. A mixture too rich at slow speeds should be corrected by cutting down on the gasoline, and at high speeds by increasing the auxiliary air. An auxiliary air spring which sticks, a restricted air opening, or a flooded carburetor will cause an overrich mixture.

(b) *Mixture too Weak.*—A weak mixture can be detected by back-firing through the carburetor and by occasional muffler explosions. A weak mixture, being a slow burning mixture, is still burning when the intake valve opens for the following charge. This permits the flame to shoot back through the manifold into the carburetor. A weak mixture should not be confused with an improperly timed intake valve which opens before the burning charge has been exhausted. If the intake valve has a weak spring which does not close the valve properly, it may permit back-firing through the carburetor. The explosions caused by the valve trouble are usually more violent than a back-fire due to weak mixture. A weak mixture at low speeds is caused generally by too little gasoline and at high speeds by too much auxiliary air and the carburetor should be adjusted accordingly.

An air leak in the manifold connections will dilute the mixture with air and cause a weak mixture and back-firing. These leaks should be remedied before the carburetor adjustments are changed.

A stuck or bent or obstructed gasoline needle valve may cause a weak mixture by shutting off the supply of gasoline. The remedy is obvious.

(c) *Color of Explosive Flame.*—By opening the priming cocks on the cylinders, the color of the flame can be seen as the explosive flame issues out of the cocks. A blue flame indicates a perfect mixture, a red flame indicates an excess of gasoline, and a white flame indicates an excess of air.

(d) *Flooded Carburetor.*—If the carburetor float becomes gasoline soaked or filled with gasoline, it will not shut off the gasoline float valve and the carburetor float chamber will become filled with gasoline. The remedy is to take the float out and if it is made of cork, have it dried out, painted with shellac and baked. If of the hollow metal type, have the float emptied and the hole soldered. A small particle of dirt under the float valve will also cause the carburetor to become flooded.

(e) *Flooded Cylinder.*—If the engine has been cranked for some little time and too much gasoline has been sucked into the cylinders, the cylinders become flooded with almost pure gasoline which condenses in the cold cylinders. This charge will not explode. The remedy is to open the priming cocks and crank the engine until the overrich mixture has been expelled or diluted. The priming cocks can then be closed and the engine will usually start. Flooding of the engine can also be caused by priming the cylinders with too much gasoline. It sometimes happens that a flooded engine can be started without difficulty after standing for several hours. The excess gasoline has evaporated in the meantime.

(f) *Cold Weather Starting.*—In cold weather, when the engine is stiff and the gasoline is hard to evaporate, it is necessary to inject a little warm or high test gasoline into each cylinder through the priming cocks. The carburetor may also be heated by the application of warm cloths. The priming gasoline can be heated to advantage by placing a bottle of it in a pan of hot water.

(g) *Frozen Carburetor.*—If there is water in the gasoline this water may be frozen in the carburetor. The water, being heavier than the gasoline, sinks to the bottom where it may freeze in cold weather. To remedy this trouble apply hot cloths to the parts affected. Never use a torch or flame of any sort around the carburetor.

(h) *Feed System Stopped Up.*—If, after priming, the engine starts and suddenly dies down, the gasoline supply may be exhausted, the feed pipe may be clogged, or a piece of dirt may have worked into the needle valve. If there is a supply of gasoline and the trouble is found to be due to dirt in the feed system, the feed pipe may be disconnected and the dirt blown out. A particle of dirt in the needle valve may be removed by screwing the valve shut and then opening it the proper amount. This trouble and also the one due to water in the gasoline can be prevented by straining the gasoline through a chamois skin before putting it into the main tank.

(i) *Loss of Pressure on Gasoline Tank.*—It sometimes happens that if a pressure gasoline system is used, the pressure becomes too low to force the gasoline from the main tank to the auxiliary tank. This causes a lack of fuel at the carburetor. A hand pump is usually furnished for increasing this air pressure on the tank.

If the car is equipped with a gravity feed system, the gasoline may fail to run to the carburetor when ascending a steep hill. It sometimes becomes necessary to back the car uphill, in which case the gasoline will run to the carburetor without difficulty.

(j) *Water Logged Carburetor.*—It sometimes happens that the carburetor becomes loaded with water, due to the fact that it can neither evaporate nor get out. This water prevents the gasoline from getting in. The water should be drained from the carburetor drain cock.

155. Ignition Troubles.—(a) *Locating Defective Plug.*—If one of the cylinders is missing at all speeds, the ignition is at fault. The cylinder can be located by opening the priming cocks and watching for the flame to come out. The cylinder without flame, out of which issues only a hiss, but no short report, is the one at fault. All of the plugs can be taken out of the cylinders and, with the wires attached, placed on the cylinder so that the threaded portions only are in contact. By turning the engine over, the defective plug can be detected.

(b) *Defective Plugs.*—A defective plug may be broken, oil soaked, carbonized, or the air gap between terminals too much or too little. If the plug is broken, it usually must be replaced by a new plug. A plug with a loose center electrode may sometimes be repaired. If carbonized or sooted up, the plug may readily be cleaned with a stiff brush and gasoline. Do not scrape with a knife, as it merely rubs the carbon into the surface of the porcelain.

The gap between plug terminals should be between $\frac{1}{40}$ and $\frac{1}{32}$ in. It should not be more or less than this amount for efficient ignition. A smooth dime is a good gage to use for setting this gap.

(c) *Locating a Missing Cylinder.*—If, after the plugs are found to be in good order, one or more of the cylinders miss, the ones at fault can be located by detaching the wire from the plug and holding the end about $\frac{1}{4}$ in. from the plug binding post. A missing cylinder will show no spark, and the trouble is due to a lack of secondary current in the wire to the plug. Instead of detaching the wire from the plug, the current can be short-circuited by placing the metallic part of a screw-driver in contact with the plug binding post with the tip of the screw-driver about $\frac{1}{4}$ in. from the metal of the cylinder. As before, the missing cylinder will show no spark. Lack of current at the plug may be due to defective wiring, weak or run down batteries, poor adjustment of vibrator or circuit breaker, engine out of time, and dirty or defective magneto connections.

(d) *Defective Wiring and Switches.*—If there is no current at the plug, the wiring system should be examined carefully for dirty and loose terminals, broken connections, and oil soaked and wet wiring. If the insulation has been worn off, the current is liable to be short-circuited or grounded through the engine or frame of the car. Defective or poor contacts at switches may also be the cause of no current at the plugs.

(e) *Dry Batteries.*—Weak or exhausted batteries are a common source of trouble. If the batteries are suspected, they should be tested with a small "ammeter." If any one of the dry cells shows less than 6 amp., it should be taken out and replaced with a new one. One weak cell will greatly interfere with the operation of the others in the set. Occasionally, a weak dry cell can be livened up temporarily by boring a small hole through the top and pouring in a small quantity of water, or better still, of vinegar. The effect is, however, only a temporary one.

Dry batteries should always be kept perfectly dry. If they become wet on the outside, there is a tendency for the battery to be short-circuited and exhaust itself. Especially is this true if water spills on the top of the battery between the terminals.

(f) *Storage Batteries.*—If the storage battery appears dead or shows lack of energy, it may be due to one of the following causes of trouble: (a) discharged; (b) electrolyte in the jar too low; (c) specific gravity of electrolyte too low; or (d) plates sulphated. These troubles are fully treated in the chapter on starting and lighting under the heading of Storage Batteries.

(g) *Magneto Troubles.*—If the ignition trouble has been located in the magneto side of the system and the plugs and wiring system have been found in good working order, attention should be turned to the magneto itself. The distributor plate should be thoroughly cleaned with gasoline to remove any foreign matter which may have collected after considerable use. After attending to this, it should be determined whether or not the magneto is generating current. This can be done by disconnecting the magneto cables and watching the safety spark gap while cranking the engine. If no spark appears there the trouble is in the magneto itself.

The contact points may be pitted or burned. They should be filed until they meet each other squarely. Be sure that the adjustment is properly made.

The carbon or collector brushes may be dirty or worn. They should be cleaned, or if badly worn replaced with new brushes.

It occasionally happens that the magnets become weak or demagnetized. They may possibly be placed in the magneto in the wrong position. If weak or demagnetized, they should be remagnetized before being replaced. Care should be exercised in getting the like poles of the magnets together on the same side of the magneto. Most magnets are marked with an "N" indicating the north pole.

(h) *Coil Adjustments.*—A frequent cause of no current at the plug is coil trouble, especially where a vibrating coil is used for each cylinder. The vibrator points become pitted, out of line, and burned, making good contact impossible. The tension on the vibrator spring becomes changed, permitting the coil to consume too much or too little current.

In the case of burned or pitted points, they should be filed flat with a thin smooth file, or hammered flat with a small hammer. In either case the points should be so shaped as to meet each other squarely.

If it becomes necessary to adjust the tension on the vibrators, the tension should be entirely taken off and gradually increased until the engine runs satisfactorily without missing. It is very important to have all the units adjusted alike. This can be easily done after a little experience. The most accurate method of coil adjustment is with a coil current indicator by which the amount of current consumed is measured. Coils are built to consume about $\frac{1}{2}$ amp. and the tension should be adjusted so that the current consumption of each coil is not much greater than this amount.

(i) *Defective Condenser.*—A sparking between the points of a vibrating coil is due to dirty or pitted points, loose condenser connections, or a defective condenser. If the latter, a new unit must be supplied.

(j) *Breakdown of Wires or Insulation.*—If no current is obtained in the secondary of a coil, when the vibrator is working as it should, the trouble is probably due to a broken wire inside of the coil. It sometimes happens that the binding post wires become loose from the post just inside of the coil. If only a slight spark can be obtained, the insulation on the inside wire may be broken down, thus causing a short circuit of the current. Obviously there is no remedy but to replace the coil.

(k) *Timers and Commutators.*—Trouble in the timer or commutator usually comes from oil, water, and dirt which has found its way inside of the housing, causing a short circuit. This foreign matter should be cleaned out of the timer in order to have it give good service. After a time, the contact points in the timer become worn and loose. New points should be put in and all loose parts tightened. If the lost motion becomes too great, it may be necessary to supply a new timer.

(l) *The Spark Setting.*—If the engine kicks back after cranking, the spark is too far advanced and should be retarded so that the spark does not occur until the piston has passed the dead center. The tendency of an early spark on starting is to cause the engine to start backward. Too early a spark at slow speeds will make the engine knock and will cause the car to jerk.

A retarded spark causes the engine to overheat and lose considerable of its power. There is no advantage of retarding the spark past center, even in starting. When running it should be advanced in proportion to the speed.

On cars equipped with automatic spark advance, the troubles due to early and late spark are not experienced. Preignition from other causes, however, may occur with either type of spark advance.

(m) *Premature Ignition.*—Premature ignition is caused by particles of carbon, sharp corners, etc., becoming incandescent from the heat of explosion and igniting the charge on the compression stroke before the spark occurs. Premature ignition occurs generally when the engine has been loaded quite heavily at a slow speed, as when going up a steep hill on high speed. Any engine will have premature ignition if it becomes excessively hot under low speed and heavy load, but the tendency to preignite is much more marked if the cylinder is full of carbon deposits. These carbon deposits should be cleaned out as explained before.

156. Lubricating and Cooling Troubles.—(a) *Engine Lubrication.*—The usual lubricating troubles are those due to the use of the wrong kind of lubricating oil or too much or too little of it. An engine with loose fitting pistons requires a heavier oil than one with tight fitting pistons, and an air-cooled engine usually requires a heavier oil than a water-cooled engine. It is very essential that a true gas engine cylinder oil be used for cylinder lubrication because it alone satisfies the requirements. Poor lubricating oil is expensive at any price and it is good economy to use the best cylinder oil obtainable. In this matter the recommendations of the manufacturer should be followed out.

An excess of lubricating oil shows itself by a white bluish smoke coming from the muffler. In addition to this, an excess of lubricating oil causes the formation of a pasty carbon deposit in the cylinder, which causes the engine to overheat.

The important things to look after are to be sure that there is a sufficient supply of oil and that the oil pump is in working order. The crank case should be drained and washed out with kerosene and new oil put in every 1000 miles.

(b) *Poor Circulation.*—Poor circulation in the cooling system is one of the common sources of trouble and when neglected is liable to give the motorist many uneasy moments. The water system must be kept filled with water. This is of especial importance in the thermo-syphon system, in which the water level must at all times be above the return pipe from the engine to the radiator in order to have the circulation continue.

A worn pump may cause poor circulation, because in most cases the thermo-syphon effect in a forced system of circulation is not enough to keep the water moving at the proper rate.

Sediment in the radiator and scale in the engine jacket may seriously interfere with the circulation of the water. Such clogging of the system comes from the continual heating and cooling of the impure water used. This emphasizes the desirability of using pure water or rain water in the radiator. The sediment and hard scale may be removed as follows: Open the drain cock in the bottom of the radiator and introduce the end of a hose in the filler of the radiator. Run the motor for about 15 minutes and the fresh water from the hose will clean out the loose sediment or scale in the water jackets and radiator. Through this process, a supply of fresh water is constantly entering the system and passing through the water jackets while the motor is running.

Next, dissolve as much ordinary washing soda as can be dissolved in enough water to fill the radiator. Then run the motor with a retarded spark until the water is brought up to the boiling point. Allow this solution to remain in the motor and radiator for several hours, after which again open the drain cock and, with a hose, again flush out the entire system with fresh water as before. In extreme cases it would be well to repeat this process several times. The final operation of flushing out with fresh water should be thoroughly done. If any of the washing soda solution is left in the motor or radiator, it may result in undesirable chemical action.

When rubber hose forms a part of the circulating system, a kink or twist in the hose may possibly cause poor circulation of the water. The inside fibers of the hose also tend to come loose and clog the system.

In the case of thermo-syphon cooling systems or in air-cooled motors, the operation of the fan is essential to the successful operation of the cooling system. If the fan belt breaks or slips, or the fan blades are bent, the air circulation through the radiator is interfered with and consequently the water is not properly cooled.

The attention which must be given to the cooling system in winter to prevent freezing has been thoroughly taken up in Chap. V. One thing to be watched in winter running is the temperature of the water. If the weather is excessively cold, the water may be cooled below the efficient running temperature of from 180° to 200°. In this case, the radiator front should be partially covered in order to keep out a part of the cold air. This will also keep the water warm for a longer time when the car is standing.

157. Starting and Lighting Troubles.—The troubles ordinarily experienced with the starting and lighting systems are taken up in the chapter treating of those subjects.

158. Transmission Troubles.—(a) *Clutch Slips.*—Clutch troubles are about the same in either the cone, plate, or multiple-disc types. The clutch either slips, engages harshly, grabs, or refuses to release. If it slips, the full power of the engine is not transmitted and the clutch becomes hot from the friction. In the cone and dry-plate types, a coating of oil on the facings will cause slipping. The wear of the facing or weak or broken springs will cause the same results. If the slipping is caused by grease and dirt, the clutch leather should be thoroughly cleaned with a rag dipped in kerosene.

(b) *Clutch Grabs.*—If the clutch engages harshly or grabs suddenly, it may be due to the drying out or hardening of the clutch leathers. A dressing of the facing with neatsfoot oil or castor oil will make it soft and permit gradual engagement. If the clutch springs are too tight, the clutch will "drag" and burn the leather facing.

If a multiple-disc or plate clutch is designed to work in an oil bath, it will engage harshly or grab if the plates become dry. The clutch will also fail to disengage when the pedal is pressed down.

(c) *Change Gears Stick.*—If the change gears stick when attempt is made to shift from one gear to another, the shifting members may be stuck on the shaft. If the gears have become burned or teeth broken out, the particles of metal may prevent the movement of the sliding member. Occasionally the shifting lever becomes stuck and refuses to operate the gears. Under ordinary conditions, the change gears should give very little trouble if due attention is given to the lubrication and care to their shifting in operation.

(*d*) *Differential Troubles.*—A noisy differential and driving gear is due to dirt, lack of grease, or broken or worn teeth. In some cases wear can be taken up by the proper adjustments, but these should always be made by an experienced mechanic. The differential, as a rule, will give very little trouble. A break in the differential or in its connections to the wheels is made evident by failure of the engine to propel the car. If the connection to either wheel is broken the other wheel will also lose its power.

159. Chassis Troubles.—(*a*) *Faulty Alignment of Front Wheels.*—Most of the front wheel trouble is due to faulty alignment. The following instructions are given for the adjustment of the front wheels and bearings on the Overland car: The front wheels, when correctly aligned, are not exactly parallel, but "toed-in" (Fig. 242). To test their proper alignment, jack up both front wheels and with a piece of chalk or a lead pencil

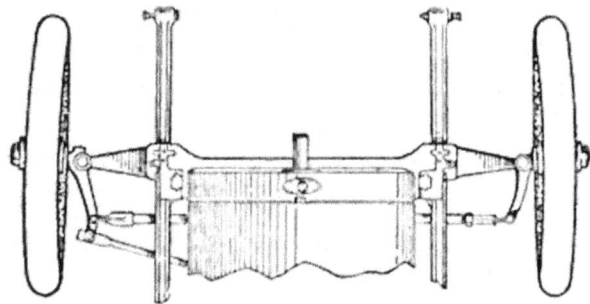

Fig. 242.—Toed-in and cambered front wheels.

held in a fixed position against the tire spin the wheels, drawing a line around the tire casing. The distance between the lines measured at the front of the wheels should be from $3/8$ to $1/2$ in. less than in the rear.

"If a steering knuckle is bent, it is best to replace it with a new one, because bending it cold will not always restore its correct shape, while heating it may make it too soft for safety.

"If faulty alignment is due to a bent steering cross-rod, it may be straightened out and then adjusted by loosening the lock-nut and screwing the rod in or out of its yoke end. Be sure to lock the nut tightly after adjusting.

"The front wheels are also 'set,' or 'cambered,' so that the wheels are a little closer together at the bottom than at the top. This arrangement is desirable on account of the fact that the front wheels are 'dished' so as to make the wheel a sort of flattened cone. This 'dish' of the wheel is compensated by the 'camber,' by which means the lowest wheel spoke is in a vertical position with relation to the road surface. The combined 'toeing-in' and 'cambering' makes for greater strength and also reduces materially the effort required in steering the vehicle. The camber is secured by inclining the axle spindle from its central line, and no adjustment is required in connection with it.

"To see whether the front wheel bearings need adjustment, jack up the wheels. Any looseness will show on rocking the wheels sideways. To tighten the bearing, spin the wheel, at the same time screwing down the adjusting nut until the bearing is so tight that it will stop the rotation of the wheel. Then back off the nut only enough to allow the wheel to spin. Lock in this position and the bearing will give the best service.

"In general, a somewhat loose bearing is to be preferred to one that is so tight that the rollers are likely to become injured."

(b) *Loose Steering Gear.*—With continued use, the worm or screw in the steering gear will wear, and a looseness of the wheel will result. Means are usually provided for taking up this wear. Most drivers prefer to have a small amount of lost motion (about $\frac{1}{2}$ in.) in the wheel, as it makes steering easier and relieves the steering gear from all the road shocks. A great deal of steering gear trouble and wear can be avoided by oiling all the joints regularly. This important point is too often neglected.

(c) *Brakes.*—It is very necessary that the brakes be kept in perfect working order at all times. It is more necessary to be able to stop the car in emergencies than to start it. If the brakes fail to hold, it may be that the drum and band facings have become covered with oil and dirt, or the band facings may be worn. In the latter case, new facings are necessary in most cases, but adjustments can be made for slight wear.

The brakes may bind or stick, due to the tight adjustments. With tight adjustments, the motor is pulling the car against the friction of the brakes at all times.

If the brakes are not adjusted the same on each side of the car, there will be a tendency for the car to skid when the brakes are applied. The braking effect comes on only one wheel and this tends to swing the car around. Many cars are provided with brake equalizers which allow them to work together.

(d) *Springs.*—After a car has been run for some little time, the spring clips become loose and the conditions are then ideal for breaking the springs. Spring breakage occurs mostly with loose clips. Consequently these clips should be tightened every once in a while.

When springs are not lubricated, water works its way in between the leaves and causes them to rust, often to such an extent that they become almost like solid pieces. This causes them to lose much of their spring action. It is a good plan to jack up the frame of the car occasionally, so as to take the weight off the springs, and insert oil and graphite between the leaves. It is also a good plan, about once a year, to have all the springs taken apart, the surfaces thoroughly cleaned and coated with a thick mixture of oil and graphite.

CHAPTER X

OPERATION AND CARE

160. Preparations for Starting.—Before starting an automobile engine, the driver should make sure that there is plenty of gasoline in the tank and that it is turned on so as to flow to the carburetor. The radiator should be filled with clean water, free from lime or other form of matter that will have a tendency to coat the inside of the radiator when the water evaporates and thus prevent cooling action. Rain water is best. The driver should also be sure that he has plenty of lubricating oil. In starting the engine, close the switch on the battery circuit, or, in some cases, where a high tension magneto is used, the engine may be started on the magneto. It is better, though, in most cases, to use the battery circuit, as the current there is always available. The change speed lever should be in the neutral position. If the lever is so that the gears are meshed, cranking the engine would start the car in motion, and engines that pick up easily are liable to start and run away, especially if the gear shift lever is in the first position. It is also advisable to have the emergency brake set. This will quite often prevent runaways. The spark lever should be *retarded*, and the throttle lever slightly advanced before cranking the engine. As soon as the motor starts, advance the spark lever about two-thirds of the distance around the quadrant, and retard the throttle lever so that the motor will not race.

161. Cranking.—In cranking the engine, always set the crank so as to pull up. In this manner, should there be a back-fire the crank will be pulled down out of the hand; whereas, if one is pushing down on the crank, the back-fire will be very liable to cause injury to the driver's wrist or arm, as he would be unable to get away from it.

After an engine has been standing for some time, it is quite probable that it will not get gasoline at once, due to the gasoline evaporating or leaking from the carburetor. In order to have sufficient gasoline in the mixing chamber, it is customary to raise the float, which allows the gasoline to overflow into the mixing chamber. This process is commonly called "priming" or "tickling" the carburetor and insures a rich mixture in starting.

This may also be accomplished by opening the priming cocks on the cylinders and pouring a few drops of gasoline directly into the cylinders. If there are no priming cocks on the cylinders, one can use a priming spark plug.

162. How to Drive.—There is "good form" and "bad form" in driving a car the same as in doing anything else. One-half the pleasure of motoring comes from knowing how to drive easily. Proper driving also means minimum strain and wear on the car. It prevents unnecessary stress and wear on the motor and transmission system, and saves the gasoline and oil. In starting the automobile, the object is to have the car pass from a stationary position into rapid motion with the least amount of stress on the motor and transmission, and also with the most comfort to the occupants of the car. In doing this, a steady pull should be maintained on the driving mechanism from the point where the driver lets in the first speed until the car is under full headway. Starting with a jerk,

Fig. 243.—Shifting gears.

or passing unevenly from one speed to another, strains the motor, racks the frame, and causes various troubles in the driving mechanism. Having started the engine with the gears in the neutral position, the proper method of gear shifting is as follows:

Advance the spark lever about two-thirds of the way around the quadrant, throw out the clutch, and throw the speed change lever in the first position, as shown in Fig. 243. Let the clutch in easily but firmly and increase the motor speed gradually, either by the foot accelerator or by the hand throttle, until the motor picks up the load. Try to accelerate the engine as the clutch is let in. The mechanical act of shifting gears is very simple, but the knack of learning to perform the operation rightly takes practice. As you engage the gears for any speed and begin to let in the clutch, give the motor more gas *at the same time*. Once you have learned to do this properly, you will never have to give it a thought.

OPERATION AND CARE

In changing from first to second speed, release your foot accelerator or throttle hand lever, then throw out the clutch, change to second speed, and again let in the clutch, at the same time accelerating the engine again. Repeat the same operation on going into higher speed.

Just before shifting gears, the engine should be throttled by removing the foot from the accelerator, so that the two gears which are going to be meshed are running at the same speed. This permits a smooth shifting of gears, and also prevents the motor from racing. Then as the clutch is let in the engine should be accelerated to give it sufficient power.

When the car is in high speed, assume a comfortable easy position. Do not sit sideways in the seat nor take your hands from the steering wheel. If one sits in an easy upright position, driving does not become tiresome, and it also gives a person better control, as he does not have to move from his position in order to operate any of the levers. Also, an erect and alert driver makes a better appearance than one who slouches in his seat and handles his car carelessly.

Fig. 244.—Emergency stop.

163. Use of the Brakes.—The operation of stopping a car smoothly is just as important as knowing how to start. The best results are obtained by beginning to pull up your car early enough, so as to apply your brakes gradually, thus bringing the car to a stop without straining the mechanism or jolting the passengers. Do not wait until you are within a few feet of the stopping place and then have to use the emergency brake or jam the brakes down hard. Applying the brakes hard is not only an unnecessary strain on the mechanism, but is very hard on tires since, when the wheels stop, the road acts as a file on the tires.

Sometimes it is necessary to make an emergency or quick stop. In doing this the operator does not take time to slow down his engine, but presses both foot pedals and applies the hand emergency brake at the same time, as shown in Fig. 244. In pressing both pedals, he releases the clutch and applies the service brake, and the braking effort is further increased by the application of the emergency brake.

In descending steep hills, it is often convenient to use the engine as a brake. This can be done by closing the throttle and shutting off the spark. Then by leaving the clutch in, the car is forced to run the engine against compression without receiving any power from it. The gear shift lever may be left in either high, intermediate, or low speed. In the low speed position the engine will have more of a braking effect than in the high speed position, because it must be turned much faster for the same speed of the car. If the grade is long and steep, use the foot and emergency brakes alternately. This equalizes the wear on them.

164. Speeding.—When running a new car, do not speed it up until you are absolutely sure of your ability to drive. Furthermore, any new piece of machinery should not be run at high speed for any length of time until its bearings have had a chance to wear to a smooth fit. A few miles of racing are harder on the bearings of a car than several days of moderate driving.

165. Care in Driving.—All cars have low and intermediate gears for use in starting, hill climbing, and bad roads. A good rule to follow in shifting gears is to shift just before you need to in climbing hills. To attempt to climb every hill on high speed always marks the amateur driver. The intermediate gears should be used on steep hills, even if they could be climbed on high speed. If it is desired to climb a hill on high speed, one should take a running start and rush up the hill. In going over bad roads, it is better to shift into second or first speeds immediately. This will save slipping the clutch, which is a bad practice. On the lower speeds, one can control the speed of the car entirely by the use of the throttle.

In going over bridges, cross-walks, railroad tracks, or water-brakes, it is better to strike them at an angle than to hit them squarely. This method throws the strain on the springs successively instead of all at once and reduces the rebound of the car. In going through sand, it is better to let the car pick its way and not try to hold it in line and force it to make a new track. For this reason a little play in the steering gear is desirable.

One of the first things that a new driver learns is the advantage to be derived from consideration and courtesy extended to others using the public highway. Most drivers know that they are expected to turn to the right when approaching a vehicle, or to the left in overtaking and passing a slow-moving vehicle going in the same direction. In meeting another car at night, dim your headlights so that they will not confuse the other driver.

After they have begun to realize the accuracy with which a car may be steered and the ease with which it may be called upon to pass another vehicle, possibly approaching from the opposite direction, it seems natural for some drivers to display their nerve in not turning from the center of the road until they are almost upon the approaching vehicle. Often, however, the other fellow has as much courage and takes the same stand, and in the confusion which very frequently follows, either one or both cars are damaged on account of collision.

In passing vehicles which are approaching, as large a margin of space as possible should be afforded, and in passing a slow-moving vehicle ahead, pass it as quickly as possible and without cutting in short ahead of it.

166. Driving in City Traffic.—The lack of consideration on the part of a few careless drivers has resulted in the adoption of very strict municipal regulation governing traffic. Those who are familiar with city traffic regulations and apply them as well on country roads, will not be likely to encounter difficulties.

The burning of at least three lamps, including two head or side and one tail lamp, is enforced from sun-down to sun-up in practically every state.

Fig. 245.—Turning to the right.   Fig. 246.—Turning to the left.

In approaching an intersection, either in the city or in the country, where a clear vision of the road approached can not be had because of buildings, fences, etc., which obstruct the view, the car should be slowed down to a speed at which it can be readily stopped in case of the approach of another vehicle from either side.

In turning into another road to the right, the driver should keep his car as near the right-hand curb as practicable, as shown in Fig. 245.

In turning into another road to the left he should turn around the center of the two and as in Fig. 246. No vehicle should be slowed or stopped without the driver thereof giving those behind him warning of his intentions to so do, by proper signals.

Often drivers of horse-drawn vehicles become confused if their horses are frightened by the approach of an automobile and in drawing up the horses sharply to one side the animals are liable to jump or rear, with the result that the vehicle may be overturned and the automobile injured as well. In cases of this kind, it is better to stop the machine entirely and, if necessary, even stop the motor.

More accidents result from unwillingness to change gears than from almost any other cause. Most American drivers use their first and second speeds only in starting their car. They allow the car to drift along and thus get into a tight place in traffic or too close to street cars and, because of misjudging the speed of the approaching vehicle or their selfish desire to crowd out another car, collisions or other accidents frequently result. It is a simple operation to change from third to second speed. It increases the power and affords the possibility of a great deal quicker acceleration as well. The second speed is incorporated for a purpose. It is seldom that we are in such a hurry that we can not spare a moment to afford absolute safety.

Accidents are not due to one's losing control of the car in many instances, but are more likely due to one's losing control of himself. One is not an expert driver until he intuitively performs the operations which control the car just as one walks or reaches out for an object.

167. Skidding.—When traveling on slippery roads, avoid making sudden turns; also avoid sudden application of the brakes or sudden changes of power, as they all tend to cause skidding.

Most skids can be corrected by the manipulation of the steering and brakes. An expert driver can keep his car straight under almost any conditions, but it is impossible to explain just how he does it, except that he knows his car and becomes almost a part of it. Usually the rear end skids first, and in the right hand direction, this being caused by the crown of the road. Under such conditions, the skidding action will be aggravated if the brakes are applied, and the car may be ditched or continue to skid until it hits the curb.

The correct action in an emergency of this kind is to let up on the accelerator pedal and thus to reduce the power to a point where the wheels are rolling freely without either being retarded by the brakes or drawn ahead by the engine. If the car recovers its traction, the power may be applied gradually and it will be advisable to steer for the center of the road again. However, if the car continues to skid sideways, steer for the center of the road, applying the power gently. This will aggravate the skid for the moment, but will leave you with the front wheels in the center of the road and the car pointing at an angle. By so doing, you can mount to the crown of the road again and the momentum of the car will take the rear wheels out of the ditch on the right hand side. It is customary to advise turning the front wheels in the direction that the car is skidding in order to correct the action, but this can hardly be said to be advisable in most cases, as the amount of room on the skidding side is somewhat limited, and for this reason the explanation given above will better apply to such a condition.

When turning a corner on wet asphalt pavements it frequently occurs that the front wheels skid. In a case of this kind, immediate action is necessary. It will be found that by applying the brakes suddenly for a moment so as to lock the wheels, the rear end of the car will skid in the direction in which the car is to be turned. This will help the action of the front wheels and the releasing of the brakes and the touch of the accelerator will bring the car around the corner without any over-travel of the front end. By applying the brakes in this way, it is possible to turn the front wheels in the direction opposite to that which the car is to be turned for a moment while the rear end is skidding. When the brakes are released, it is plain to see that the front wheels will have no tendency to skid farther, as they will be pointing in the direction which the car is to be turned and the rear end will be in line with it, due to the skid.

Needless to say, this manipulation requires a little more expertness than the correction of an ordinary skid on a straight road.

Skidding can be prevented and accidents avoided, also the life of the tires lengthened, if one will learn how to turn his car out of street car tracks and ruts. Make a sharp turn of the front wheels. Do not allow the wheel to climb along the edge of the rut and finally jump off suddenly, and do not attempt to climb out of these conditions at speed.

Driving a car around a sharp corner at 25 miles an hour does more damage to the tires than 15 or 20 miles of straight road work. This is an economical reason why one should drive around corners cautiously and slowly. The other reasons are obvious.

The natural inclination of the driver is to throw out the clutch in coasting down hill or driving over rough roads. This should not be done. Keep the motor pulling the car over rough roads. Thus it keeps everything taut and lessens the shock and jar that the car gets through bumping over ruts.

168. Knowing the Car.—One will very soon become accustomed to all of the noises the car makes, and any strange sound, be it ever so slight, will be immediately perceptible.

Much of the satisfaction that an automobile gives depends upon the driver. If he neglects his automobile, if he does not lubricate it, or if he tinkers with it too much, he is bound to receive unsatisfactory service.

No machine can be absolutely automatic. All things must wear in time. The best preventive of wear, and the most certain thing to increase the life of an automobile, is proper lubrication. Remember that a motor car is like any piece of machinery and will not keep in good running condition without a reasonable amount of care. The life of a car can be cut in two by neglect or doubled by careful use.

One should familiarize himself thoroughly with all the lubricating points of the car. The chart in Chap. V will show where each one is located. Make the lubrication of the car as regular as the eating of meals. If one does this he will not have any complaint to make of his car becoming noisy or of bearings wearing out. If he does not do it, he will not get the satisfaction from his car that he expects. Satisfaction would be greatly increased if everyone would learn the details of his machine, that is, learn to make the simple examinations and adjustments. Do not depend on some one else to do that which is so simply done and which one can get much satisfaction in doing. One should familiarize himself with every detail of his car and then he will have great confidence in venturing over any road at any distance from a repair station.

In learning to drive a car, it is better to use the hand throttle for the first few days until you have mastered the other details of driving. Then learn the use of the foot accelerator. The foot accelerator is controlled by a spring and is released by removing the foot. This will slow down the car to the point where the hand throttle is set. In using the foot accelerator, keep the hand throttle set at a point where the engine will just pull the car. Then, when the foot is removed from the accelerator, there will be no danger of an accident from the car's not slowing down.

Never allow the motor to race when it is idle. When there is no load on the engine it will vibrate unduly at high speeds, which causes excessive strains and makes the engine and car noisy. Racing the motor when driving can be avoided by learning to use the foot accelerator in the proper manner in relation to the clutch and gear shifts.

169. The Spring Overhauling.—The greatest trouble with the average motorist is that he has the idea that all the attention a car needs is to keep it full of gasoline, oil, and water. There are many owners, however, who enjoy making their own adjustments and keeping their car always in good condition by giving it frequent attention. After a car has been laid up for some time the oil is forced out of the bearings and, if run in this condition, considerable damage is liable to result. All old oil should be drained off and the case thoroughly washed out with kerosene. Hot kerosene and oil should be poured into the cylinders to cut the gummed oil and to remove any rust that may have formed. After draining off the kerosene, the crank case should be filled with oil to the upper test cock. Do not use the electric starter until you are sure that the motor is free to turn. Better turn the motor over a few times with the hand crank first. Clean the spark plugs by washing with gasoline and a

OPERATION AND CARE

brush—never scrape them, then adjust the spark gap between points to about $\frac{1}{32}$ in. or the thickness of a well worn dime.

Test for leaks around the valves and spark plugs by squirting oil on the joints and then turning the engine over. If there are any leaks, air bubbles will be seen in the oil.

If the gasoline does not flow to the carburetor, remove the feed pipe and blow it out; also clean the screen in the bottom of the carburetor. The gasoline flow can be tested by holding down the float.

In the wet type multiple-disc clutches, the oil should be drained off and then they should be filled with kerosene. Replace the plug and start up the motor. Let the motor run for a few minutes during which time push the clutch in and out several times. Then stop the motor, drain off the kerosene, and fill with the proper amount of lubricant. The transmission, differential, and universal joint should also be washed out and repacked. Every point mentioned on the lubrication chart of Chap. V should be cleaned, adjusted and oiled.

Electrical System.—Remove the rotor and clean its bearings with gasoline and a cloth, then rub a little vaseline on the race very lightly. Clean the breaker points with a fine piece of emery cloth and set the gap to the width of the gauge, or about $\frac{1}{64}$ in. See that all wiring connections are tight and free from corrosion. It is a good plan also to put in new dry cells and be sure that they are connected up properly.

The storage battery is probably the most delicate part of the car and should receive very careful attention. It is advisable to give the battery a long overcharge at the beginning of the season, especially if the car has been laid up for some time.

During the out-of-season period, rust will accumulate in the radiator and engine jacket, and should be cleaned out. To do this, drain out the anti-freezing solution and fill the radiator with a solution of soda and water. With this solution in the cooling system, run the motor for about 10 minutes and wash out the system, following the instructions of Art. 156(b), Chap. IX.

The leaves of the springs should be spread apart and a mixture of oil and graphite inserted.

If the tires have been removed for storage, see that a thorough application of soapstone is applied to the inside of the rims to prevent their sticking to the tires.

An easy way to calculate pressure for tires is to multiply the diameter of the tire in inches by 20. For example, the correct pressure for a 3-in. tire is 60 lb., and for a 4-in. tire, 80 lb. A tire should be pumped up till it becomes perfectly round when supporting the weight of the car. Of course the only sure way of getting the correct pressure is with the use of a reliable pressure gauge.

170. Washing the Car.—The car should be washed before the mud has a chance to dry. If a hose is used, the stream should be tempered or, better still, the nozzle should be taken off the hose and a slow stream used. Always use cold water, as warm water will injure the varnish. After hosing off the mud, take a sponge well filled with water and gently dash it against the surface. Never rub the surface when washing, as it is sure to scratch the polished surface.

After the mud has been removed, remove any grease from the finish by washing with suds of a pure white soap. This should be done with a soft sponge and as little rubbing as possible. After soaping, rinse with cold water, rub dry, and polish with a chamois skin. Do not have the car standing in the bright sunlight, for it will dry too rapidly and be streaked.

A new car should be washed with *cold* water before it gets dirty. The cold water will help to set the varnish and prevent the accumulation of dust.

Cleaning the Reflectors.—When lamp reflectors become dirty do not wipe them, but use a stream of cold water to remove the dust or dirt and permit the reflectors to dry by air only. The reflectors are silver plated. The silver becomes scratched when the reflector is wiped, even with very soft material. If reflectors become dull after long service, they should be polished by using chamois with a light application of red rouge or crocus. The chamois should be very soft and free from wrinkles. If a wad of cotton or waste (about the size of an egg) is placed within the chamois, a smooth surface for wiping can be obtained. Red rouge or crocus is used by jewelers for cleaning watch-cases. When properly placed on chamois, it will not scratch the reflector. Moisten the chamois with alcohol, then apply the rouge or crocus to the chamois and wipe the reflector with a continuous rotary motion, but do not press too hard. The polishing marks will be very noticeable if other than a rotary motion is used. The efficiency of old reflectors will be increased if they are silver plated. This should be done by a lamp manufacturer or a reliable silver-plater.

171. Care of Tires.—The following few suggestions will apply to pneumatic tires in general. The various sizes of tires are constructed for the purpose of carrying up to certain maximum loads and no more. Owners should realize, therefore, that overloading a car beyond the intended carrying capacity of the tires is sure to materially shorten their life.

Do not turn corners or run over sharp obstructions, like car tracks, at a high rate of speed. Such practice is sure to strain or possibly break the fabric, with the result that the further life of the tires will be limited. Remember that most tire troubles are the result of abuse more than use.

In case of puncture the car should be stopped at once and the tube repaired or replaced. The tire should also be examined carefully and the cause of the puncture ascertained, and the nail, glass, or whatever it may be, should be extracted. Before replacing the tire on the wheel, examine the inside of the casing to see that the cause of the puncture is not still protruding, because, if allowed to remain, it would continue to cut the inner tube. It is also advisable to look over the outside of your tires frequently and take out any pieces of glass or other particles which may have become imbedded in the casing, as they are liable to work themselves in and finally puncture the inner tube.

A puncture, gash, or cut sufficiently deep to expose the fabric should have a vulcanized repair made without delay. Otherwise, water and dirt will soon ruin the whole tire, the threads acting as a conductor for the moisture, the fabric thus becoming rotted.

A bruise is an injury to the carcass of a tire caused by violent contact with an irregularity which tears the fabric. Usually the injury does not show at once. However, the structure of the tire is permanently weakened at the injured spot, and eventually a blowout will occur. Even the most careful and skillful driver cannot avoid bruises altogether. But if your tires are properly inflated and you strike an obstruction, the tire has the resiliency of the air behind it to aid in resisting the impact of the blow and the effect is likely to be less serious.

Experience has taught the careful driver to carry one or more spare tubes, as a cemented roadside repair will not always hold, especially in warm weather, as the heat generated in the tire may loosen the patch. When touring, a spare casing should always be carried. It should be strapped tightly to the tire holder, otherwise it will chafe.

Spare tubes should be kept lightly inflated. This keeps them in good condition and prolongs their life. They should not be stored in a greasy tool-box under any circumstances.

Excessive weight on a casing will break down the fabric in the side walls, and if persisted in, a blow-out is apt to result. When this occurs, the casing is likely to be so badly damaged as to be beyond repair. If your roads are very rough and stony, or if you are carrying heavy weights in your car, it is better to equip the car with a set of extra-size tires. You can get larger tires which will fit your rims.

Pneumatic tires are designed to carry loads in proportion to their cross-sectional area and diameter. They should never be overloaded. Following is given a table of the various tire sizes and the weight each tire should carry. Weigh the car, and if the tires are carrying more than their rated load put on larger tires.

Size of tires	Load per wheel in pounds	Size of tires	Load per wheel in pounds
2½ in. all diam.	225	30 × 4 in.	550
3 in. all diam.	350	32 × 4 in.	650
28 × 3½ in.	400	34 × 4 in.	700
30 × 3½ in.	450	36 × 4 in.	750
32 × 3½ in.	550	32 × 4½ in.	800
34 × 3½ in.	600	34 × 4½ in.	900
36 × 3½ in.	600	36 × 4½ in.	1000
		All 5 in.	1000 or over.

If the car is not used during the winter, it is better to remove the tires from the rims, keeping casings and tubes in a fairly warm atmosphere away from the light. It will be better to slightly inflate the tubes, as that keeps them very nearly in the position in which they will be used later on. Before the tires are put back, the rim should be thoroughly cleaned and any rust carefully removed; a coat of paint or shellac is also advised.

If the tires are not removed and the car is stored in a light place, it will be well to cover the tires to protect them from the strong light, which has a deteriorating effect on rubber.

The greatest injury that can be done to tires on a car stored for the winter is to allow the weight of the car to rest on the tires. The car should be blocked up, so that no weight is borne by the tires, and the tires should then be deflated partially. This will relieve the tires of all strain, so that in the spring they should be no worse for the winter's storage.

Extra casings carried on the car should be covered to protect them from the sunlight, which has an injurious effect on rubber. Do not place your extra tubes where they will come into contact with tools or oil. Carry the tubes in a tube bag. It is a good plan to tie a piece of cloth around the valve stem before placing the tube in the bag. This will prevent the possibility of the stem injuring the rubber.

Bear in mind that heat, light, and oil are natural enemies of rubber. When grease comes into contact with your tires, it should be removed immediately with gasoline.

Fast driving and tire economy have absolutely nothing in common. High speed and high bills for tire maintenance usually go hand in hand. It stands to reason that the wear and tear on tires is far greater when a car is driven at a high rate of speed than when it is used at a moderate pace. In addition to the increased force with which a wheel strikes an obstruction, when rolling at an excessive speed, fast driving generates increased heat in your tires, causing disintegration.

Shifting Tires.—Tires that show wear on one side from use on rutty roads or from driving in car tracks should be turned around. It is also a good plan to place the rear tires on the front wheels when they begin to show age. Rear tires carry more than half the weight of the car, get the roughest usage, and are also the driving tires, so that they naturally wear more rapidly than the front tires, which are simply subject to a rolling action and usually sustain less weight. A sprung axle will often cause quick wearing of a tire, for the reason that the tire is running at an angle with the direction of the car. This necessarily sets up a sliding and scraping on the road surface. If the surface of one tire looks as if it has been sandpapered, examine the alignment of the wheels.

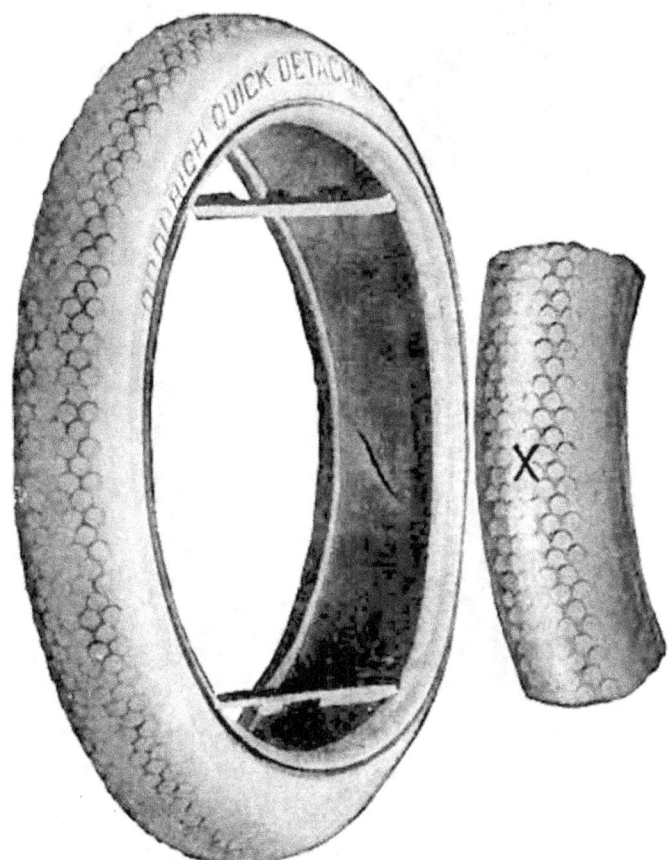

Fig. 247.—Broken fabric.

172. Tire Troubles.—*Broken Fabric.*—On the inside of the casing shown in Fig. 247 will be noticed a break in the fabric. This is the result of the blow received by the tire in hitting a stone, rail, or something of that sort at high speed. While no permanent mark may be left on the outside of the tire, especially if the object is smooth and blunt, the fabric inside may give way under the abnormal strain of such a blow. This does not indicate that the tire was in any way defective.

Sometimes a tire may be run weeks after the fabric is broken from the bruise before the blowout occurs. It has even happened in a garage, with the car standing still. Sometimes the break will exist only in a few of the plies of fabric, which will pinch the inner tube, allowing the tire to deflate gradually.

Blowouts.—Few people realize the tremendous pressure tending to rupture a tire and the consequent great strength that must be given any repair that is to be effective. This is especially true in cases of blowouts. Figure 248 shows a tire that has blown out due to ineffective repairs.

Fig. 248.—Blow-out from ineffective repairs.

It originally had a small cut extending clear through the casing. An inside patch, applied by the owner, did the tire more harm than good. The result, as shown in the picture, was that the pressure forced the patch through the hole, the patch wedging the fabric apart and causing it to break almost from bead to bead. The inside view shows how the patch has been pulled away from its original position and has been forced through the break. This condition results from the tire not receiving the proper attention when first cut. An inside protection patch, used with an outside emergency band to take the strain at the weakened point, should be used until permanent repairs can be made.

Skidding.—Skidding, or sliding the wheels by too great a brake pressure, has a disastrous effect on tires. Dragging the wheels for even a short distance over a hard rough surface will grind off the tread and even go through several thicknesses of fabric. There is nothing to be gained by sliding the wheels. Learn to apply the brakes up to the point where the wheels will just turn and no farther. The braking effect will be just as great or even greater than if the wheels are skidded.

Fig. 249.—Rut-worn tire.

Fig. 250.—Tire injured by chains.

Running in Ruts.—No tire will stand the wear from continued running in car tracks or ruts.

Figure 249 shows a tire worn off on the sides, commonly called "rut-worn." The same condition will result if a tire is run on muddy roads that have a frozen crust insufficient in thickness to support the car, so that the tire in breaking through is bound to be gouged off in the manner shown. This condition also results from running close to and rubbing against curbstones. A similar condition, but nearer the tread, is caused by running in car tracks.

One can readily see that this puts the side of the tire to a greater test than its surface ever gets in merely passing over the road. No tire will withstand this rough treatment.

Chain Bruises.—Figure 250 shows a tire that has been injured by the use of chains. Almost any chain will injure a tire if used to excess, but some are more injurious than others. Evidently, the chain used on this tire was fastened to the spokes; at least, it appears that it was held tightly in one place, as the cutting appears at regular intervals. The tread is cut through the fabric and, in fact, loosened up and torn badly in places. The least injury results from chains that are loosely applied and have play enough to work themselves around the tire, distributing the strain to all points alike. The greatest amount of injury comes from using the chains on hard paved streets, where they are least needed.

Poor Alignment.—Figure 251 shows a tire that is worn to the fabric. This is a very common condition, and is caused by the wheels being run out of line and usually occurs on the front wheels, affecting both tires alike, although sometimes one tire only is affected. Improper adjustment of the steering apparatus, or a bent knuckle, cross-rod or axle is responsible. Under either of these conditions the tread will wear away in a remarkably short time.

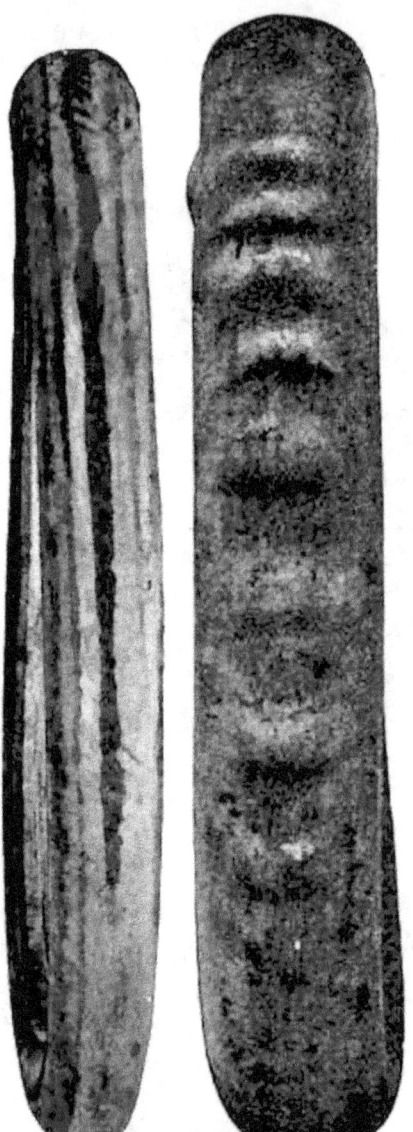

FIG. 251.—The result of poor wheel alignment. FIG. 252.—Result of under-inflation.

It is to be assumed that all cars are received from the manufacturer in perfect alignment, but after being run a while, the steering gear, if not watched very closely, is apt to become affected by wear or accident. To aid in steering, the front wheels are permitted to "toe in" just a little, but if allowed to do so to any marked degree, this condition is bound to result.

Under-inflation.—Figure 252 shows the result of running a tire under-inflated, that is, too soft. In this condition, the tire is being constantly kneaded by the road surface and the rubber is worked loose from its bond to the fabric. The wavy condition of the tread is due to this loosening. Another condition which is not visible in this figure is rim-cutting. There are probably more tires injured from this cause than any other. Proper inflation will prevent both conditions. There is a mistaken idea among many motorists that it is easier on tires if they are not inflated quite to the pressure recommended. Keep the pressures up to those recommended. There is little danger of over-inflation unless an air bottle is used. The prevailing pressure for tires is 20 lb. times the diameter of the tire. For example, the pressure for a 4-in. tire is 20 times 4, or 80 lb. Of course, the pressure should vary somewhat with the weight on each tire, but if a car is properly tired the above figures will hold. In the absence of any better test, a good rule to follow is to inflate to a sufficient pressure to prevent the tires from showing any depression under the weight of the car without passengers.

Blisters.—Small cuts in the rubber, especially if they extend to the fabric, should be given immediate attention. If these cuts are neglected, the tread will work loose from the fabric, sand will work in and form a sand blister. Furthermore, water reaches the fabric and quickly rots it so that a blow-out may soon result. As soon as discovered, such cuts should be cleaned out and the cut filled with some plastic tire compound made for this purpose.

173. Figuring Speeds.—In order to figure the speed of any automobile, it is necessary to know three things, namely: the speed of the engine in revolutions per minute, the gear ratio or gear reduction, and the size of the rear wheels. To make this figuring unnecessary the chart of Fig. 253 has been produced, from which the result can be taken without any actual figuring.

Thus, beginning at the bottom on the left hand side, the diameter of the wheels is 37 in.; follow vertically up the 37-in. line until it intersects the gear ratio diagonal. In this case the gear reduction is $3\frac{1}{4}$ to 1. The 37-in. line intersects this diagonal at the point C.

Then follow horizontally across to the right hand side of the chart, where such a horizontal line would intersect the diagonals representing the speed of the engine. In this instance the engine speed is taken at 2000 r.p.m., and the line intersects it at the point D. From this point drop a vertical to the base, which will be intersected at a point representing the car speed, in this case 67 miles per hour.

The table can also be used to find the engine speed in revolutions per minute, knowing the car speed in miles per hour (*which can be read on the speedometer*), the size of tires and the gear reduction. In such a case proceed as before, obtaining the horizontal line *C–D* extending across the diagram. Then starting on the right hand base line, at a point indicating the speed as 67 miles per hour, draw a line vertically upward until it intersects this *C–D* line. This point of intersection *D* will come on a diagonal, giving the speed of the motor. In this case it comes on the 2000-r.p.m. line exactly, but if the speed were followed upward from 50 miles per hour, for instance, another point would be obtained not on any of the curves drawn. However, it would be midway between 1600 and 1400, so that 1500 r.p.m. would be taken as the motor speed.

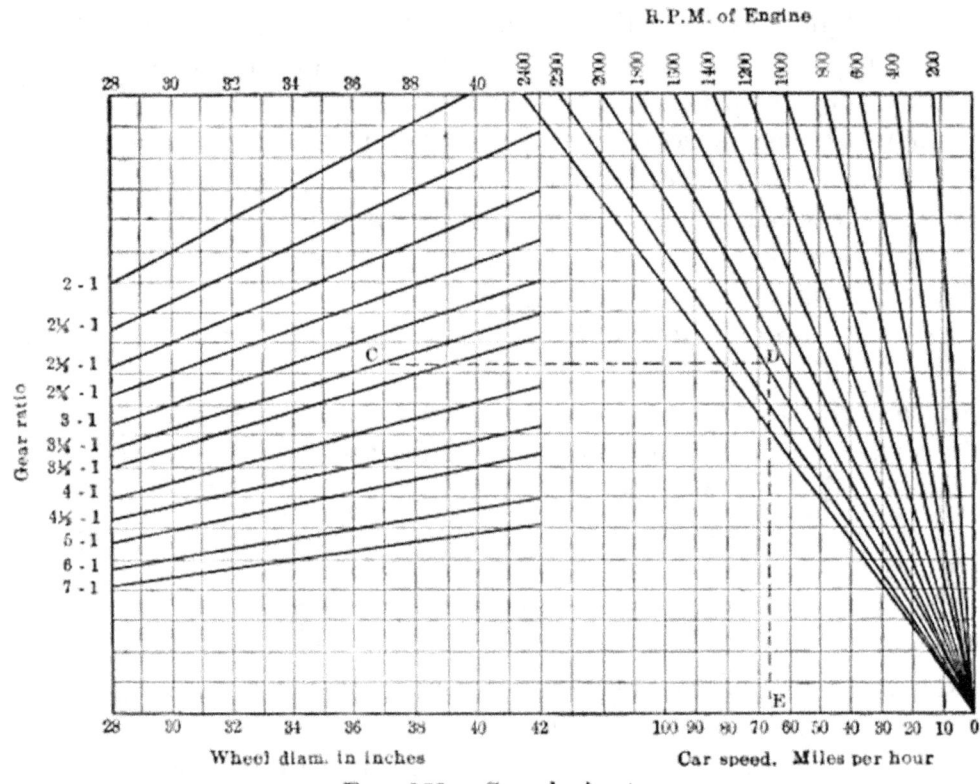

Fig. 253.—Speed chart.

174. Interstate Regulations.—The lighting requirements of the different states are practically uniform and call for two white lights in front and one red light in the rear. It is usually required that the rear license tag be illuminated with a white ray from the rear lamp. Many cities now require that the headlights be dimmed. This makes it desirable to inquire regarding such regulations before driving through a strange city.

All states with the exception of Louisiana require the registration or licensing of automobiles in some form, but the law in Mississippi has been declared unconstitutional. The registrations are renewable annually except in the District of Columbia, Florida, South Carolina, Tennessee, Texas, and Utah, where they are perpetual, and in Minnesota, where they are for 3 years. Professional chauffeurs must be examined and licensed in nearly all states, while in some states even the owner and the members of his family must have drivers' licenses.

Non-residents of a state are permitted to drive in most of the states for limited periods without taking out a license, providing they have complied with the laws of their own states and providing their own states reciprocate in this respect. These periods vary from 10 days in New Hampshire and Rhode Island to 90 days in California and Colorado and to unlimited periods of some others.

In Oklahoma, South Carolina, Tennessee and Texas, non-residents are not exempt from registration, but the fee is only from 50 cents to $3 for these states. Oklahoma also permits its cities to license and regulate the use of automobiles. In Connecticut, non-residents are permitted to travel on their home licenses only provided they have *two* license tags, one front and one rear. In Louisiana, the entire control is left to the municipalities.

In Alaska, there is no license required except for dealers. In Porto Rico, non-residents must secure a license from the Commissioner of the Interior. The fee is $2 per month.

The motorist must remember that there are local restrictions everywhere, which could not be given in the limited space available here, even if all of them were available. For instance, Wisconsin, Pennsylvania, New York City, Detroit, Chicago, Province of Ontario, etc., either require a full stop or slowing to 4 or 5 miles per hour in approaching a street car stopping to let off or take on passengers. These and local traffic police restrictions can be found out locally, or avoided entirely by driving slowly and carefully at all times, and in a manner consistent with the rights of others, particularly of pedestrians.

In case of accident, the motorist should always stop, obtain the names of witnesses, and give his own name and other information freely, as well as evidence a willingness to assist, whether in the wrong or not.

175. Canadian Regulations.—Upon entering the Dominion, the owner or operator must give a bond for the re-exportation of the car. This is to prevent cars being taken in permanently duty-free. In the majority of provinces, a Dominion license and tags are necessary.

If the tourist is not known personally to the officer at the border, he must take out the license and give the bond as mentioned above. But if known, he may be allowed to enter free of both duty and tax for 7 days.

The bond given must be for twice the amount of duty, if the stay is to be for less than 6 months. This is furnished by bonding companies in the principal cities of the United States and Canada, and usually at the border line, the usual fee being $5. The following are among those who will furnish such a bond: Guarantee Co. of North America, 111 Broadway, New York City; J. A. Newport & Co., Niagara Falls, Ontario; Niagara Falls Auto Transit Co., Niagara Falls, N. Y.; J. M. Duck, Windsor, Ontario; A. J. Chester, Sarnia, Ontario. Messrs. Newport and Duck will also procure the license and permit in advance, if requested, the charge being $4.30.

176. Touring Helps—Route Books.—The whole of the United States and the tourable parts of Canada are covered by the Automobile Blue Books. Of these there are seven volumes, as follows: Vol. 1, New York State and Lower Canada; Vol. 2, New England and the Maritime Provinces of Canada; Vol. 3, New Jersey, Pennsylvania, Delaware, Maryland, and Southeastern States; Vol. 4, The Middle West to the Mississippi River; Vol. 5, The Far West from the Mississippi to the Pacific Coast; Vol. 6, California, Oregon, Washington, British Columbia; Vol. 7, the Metropolitan Guide. They are published by the Automobile Blue Book Publishing Co., 2160 Broadway, New York, and 910 S. Mich. Ave., Chicago, at $2.50 a volume. There are also other good route books published in different localities, among which is Kings Guide, which covers the north central states in great detail. This is issued by Sidney J. King, 626 S. Clark St., Chicago.

For its members, the American Automobile Association maintains a route bureau and sells a number of excellent maps.

For those who can use them, the topographical maps of the United States Geological Survey are most accurate and very interesting, giving more detailed information than any of the others, particularly with regard to difference of elevation. Information relative to them, prices, etc., may be obtained from the Director of the Survey, Washington. In some states, county highway maps may be secured from the state highway department.

177. Cost Records.—It is always a good plan to know just what the operation of an automobile costs. The following forms are suggested for keeping data on which to base figures for the annual cost statement. These forms can be ruled on the pages of any notebook of about 5 in. by 8 in. size. The notebook should be kept in the car so that complete records can always be made. In preparing an annual statement of the cost, it is customary to charge an annual depreciation of 20 per cent of the original cost of the car. The total cost for the year should include this depreciation charge, as well as the cost of gasoline, oil, tires, fines, and repairs. Accessories are more properly chargeable against the capital account of the car less an annual depreciation charge, the same as the car itself. The cost record will also give the owner a reliable record of the service obtained from his tires and the cost per mile.

GASOLINE

Date	No. of Gal.	Cost	Speedometer readings	Notes on carburetor adjustment

Total gal.——————————— Total cost———————

Miles per gal., avg.——————— Cost per mile———————

LUBRICATING OIL

Date	Gal.	Cost	Speedometer readings	Brand of Oil

Total gal.——————————— Total cost———————

Miles per gal., avg.——————— Cost per mile———————

TIRE RECORD

Make——————— Serial No.——————————— Size———————

| Date on | Date off | Speedometer Reading | | Front or Rear |
		On	Off	

TIRE REPAIRS

Date	Nature	Cost	Remarks

SUMMARY

First Cost——————————— Total Mileage———————

Repairs——————————— Cost per Mile———————

Total Cost———————————

NOTE: Keep a separate sheet for each casing and tube.

OPERATION AND CARE

REPAIRS

Date	Name	Cost		Remarks
		Part	Labor	
Total cost				

ACCESSORIES

Date	Name and Make	Cost	Remarks

FINES

Date	Place	Amount	Remarks

INDEX

A

Air cooling, 122
Alcohol as a fuel, 78
    heating value, 79
    use in radiator, 124
Alignment of wheels, 246
Alternating current, 127
Ampere, definition of, 127
Armature of magneto, 156
Atwater Kent ignition, 141
Automatic spark advance, 151
    Atwater Kent, 143
    Delco, 151
    Eisemann, 163
    Westinghouse, 146
Axles, dead, 12
    front, 8
    live, 13, 71
    rear, 12, 71

B

Batteries, dry, 128
    storage, 128, 182, 224
Battery charging, 185
    connections, 129
    ignition, 130
    troubles, 224
Bearing troubles, 221
Bevel gear drive, 71
Bloc cylinder castings, 55
Blow-outs, tire, 244
Bodies, types of, 2
Bosch magneto, 167
    dual system, 170
    two-independent system, 173
Brakes, 16
    troubles, 230
    use of, 233
Buick oil pump, 108
    rear axle, 73
Burton process, 76

C

Cadillac cooling system, 121
    "eight" engine, 60
    "four" engine, 51
    oiling system, 111
Calcium chloride, 124
Cam angles, 30
    shafts, 58
Canadian regulations, 249
Carburetor adjustments, 98
    principles, 79
    troubles, 221
Carburetors, Carter, 97
    Holley, 86, 87
    Kingston, 90
    Marvel, 91
    Rayfield, 95
    Schebler, 82, 84
    Stewart, 89
    Stromberg, 94
    Zenith, 94
Cars, electric, 1
    gasoline, 2
    steam, 1
    types of, 2
Cells (see "Batteries")
Change gears, 66
Charging batteries, 185
Chassis, the, 2
    Ford, 48
    Hollier "eight," 47
    Mitchell "eight," 46
    Studebaker "six," 5, 45
    truck, 12
Clearance and compression, 39
Clutches, 64, 228
Clutch troubles, 228
Coils, vibrating, 132
    non-vibrating, 137, 156
Cold test for oils, 104
Commercial cars, 4
Compression, 39, 216

Condensers, 132, 225
Connecticut ignition system, 139
    magneto, 160
Carbon deposits, 220
Control systems, 23
Cooling the cylinders, 40, 117
    solutions, 123
    troubles, 227
Cost records, 250
Cranking, 231
Crank shafts, 57
Current, direct and alternating, 127
Cycles, 25
    four-stroke, 26
    two-stroke, 35
Cylinder cooling, 40, 117
    oils, 104

D

Delco ignition, 147
    starter, 190
Depreciation, 250
Differential gear, 13
Direct current, 127
Disc clutch, 65
Displacement, piston, 39
Distributor system, 137
Dixie magneto, 166
Drive, final, 70
    -shaft, 69
Driving, 232, 234
    in city, 235
Dry battery, 128
    troubles, 224
Dual ignition, 160

E

Eclipse Bendix drive, 197, 203
Eisemann magneto, 161
Electrical definitions, 127
Electric cars, 1
    ignition, 39, 127, 153
    starters, 181
Electrolyte, 184
En bloc cylinders, 55
Engine, 25
    Buda, 52
    Cadillac, 51, 60
    Ford, 53
    Franklin, 56

Engine, Jeffrey, 54
    Knight, 33
    Mitchell, 55, 63
    Packard, 63
    Speedwell, 34
    Studebaker, 52
    troubles, 214, 216
    Wisconsin, 50
Engines, eight cylinder, 60
    four cylinder, 50
    four-stroke, 26
    horse power of, 41
    six cylinder, 56
    twelve cylinder, 63
    two-stroke, 35

F

Feed systems, gasoline, 99
Fire test for oils, 104
Firing order, four cylinder, 57
    eight cylinder, 62
    six cylinder, 58
Flash point of oils, 104
Flywheels, 38
Force feed oiling, 111
Ford chassis, 48
    control, 23
    cooling system, 119
    engine, 53
    lubrication, 106
    magneto, 174
    rear axle, 72
    timer, 135
    transmission, 69
Four-stroke engine, 26
Frames, 6
Franklin, cooling, 122
    engine, 56
    frame, 6
Friction, 103
Fuels, 75

G

Gasoline, 77
    heating value of, 79
    mixtures, 79
    records, 251
Gear sets, sliding, 66
    location of, 44
    planetary, 67
Glycerine for cooling, 124

INDEX

Gravity feed system, 99
Gray and Davis starter, 193
Grinding valves, 217

H

Holley carburetors, 86, 87
Hollier "eight" chassis, 47
Horse power formulas, 41
Hydrometer, battery, 184
    Baumé, 77

I

Ignition, 39
    systems, 127, 153
    troubles, 223
Inductor magneto, 163

J

Jesco starter, 205

K

Kerosene, 78
    heating value of, 79
Kingston carburetor, 90
Knight car, Lyons, 49
    engine, 34
    oiling, 113
K-W magneto, 163
    master vibrator, 137

M

Magneto, Bosch, 167
    Connecticut, 160
    definitions, 177
    Dixie, 166
    Eisemann, 161
    Ford, 174
    K-W, 163
    Remy, 157
    troubles, 225
Magnetos, principles of, 155
    high and low tension, 156
Magnets, 153
Manifolds, intake, 102
Marvel carburetor, 91
Master vibrators, 136
Mechanism of engines, 28

Mitchell "eight" chassis, 46
    engine, 63
    "six" engine, 56
Mixtures, fuel, 79
Mixture troubles, 222
Motors (see "Engines")
    starting (see "Starters")
Mufflers, 40

O

Ohm, definition of, 127
Oiling (see "Lubrication")
Oil pumps, 106
    records, 251
Oils, cylinder, 104
Overhauling the car, 238
Overland oiling, 109
    cooling, 118
    valve adjustment, 218

P

Packard engine, 63
Parallel battery connections, 129
Petroleum, 75
Pfanstiehl coils, 133
    master vibrator, 137
Piston displacement, 39
Planetary gear set, 66
Plugs, spark, 135
Power diagrams, 43
Power, horse, 41
    plant and transmission, 14, 43
        troubles, 214
    plants, 50
Pressure feed systems, 100
Pressures, for tires, 247

R

Rayfield carburetor, 95
Rear axles, 12, 71
Records, cost, 250
Regulations, interstate, 248
    Canadian, 249
Remy battery ignition, 149
    magneto, 157
Repair records, 253
Rims, 20
Rittmann process, 76

INDEX

Rotary valves, 34
Route books, 250

S

Schebler carburetors, 82, 84
Series battery connections, 129
Shafts, cam, 58
    crank, 57
    drive, 69
    propeller, 69
Silent Knight engine, 34
Skidding, 236, 245
Spark advance, 151
    Atwater Kent, 143
    Delco, 151
    Eisemann, 162
    Westinghouse, 146
    plugs, 135
Speedometer drives, 21
Speeds, figuring, 247
Splash oiling system, 106
Springs, 6
    care of, 230
Starters, 180
    Delco, 190
    electric, 181
    Gray and Davis, 193
    Jesco, 205
    U. S. L., 204
    Wagner, 197
    Ward-Leonard, 187
    Westinghouse, 199, 200
Starting in cold weather, 222
    generator troubles, 209
    motor troubles, 208
    on spark, 179
    system, care of, 207
Steam cars, 1
Steering gear, 10
Stewart carburetor, 89
    vacuum feed system, 100
Storage batteries, 128, 181
    battery, care of, 209
      in winter, 209
      troubles, 209, 224
Stromberg carburetor, 94
Strut rods, 16
Studebaker chassis, 5, 45
    cooling, 119
    engine, 55
    gear set, 68

Studebaker ignition, 149
    oiling, 119
    starter, 199

T

Thermo-syphon cooling, 118
Three point motor support, 44
Time of spark, 151
Timers, 135
Timing, magneto, 176
Tires, 19, 240
    pressures for, 247
    records, 252
    troubles, 243
Torque arm, 15
    tube, 16
Torsion rods, 16
Transmission gears, 66
    location of, 44
    planetary, 66
    troubles, 228
Troubles, 213
Trucks, 4
Two-stroke engines, 35

U

Unisparker, 142
Universal joints, 15, 69
U. S. L. starter, 204

V

Valves, 30
    adjustment of, 217
    arrangements of, 32
    grinding, 217
    rotary, 34
    timing, 29, 219
Vaporization, principles of, 76
Viscosity of oils, 104
Volt, definition of, 127
Voltage of dry cell, 128
    of spark, 132
    of storage cell, 129

W

Wagner rectifier, 186
    starter, 197
Ward-Leonard starter, **187**

INDEX

Washing the car, 240
Water cooling systems, 117
Westinghouse ignition system, 144
    starters, 199, 200
Wheel alignment, 229, 246
Wheels, 18
Winter cooling solutions, 123

Wisconsin engines, 50
    oiling system, 112
Worm drive, 71
    steering gear, 10

Z

Zenith carburetor, 94

©2011 Periscope Film LLC
All Rights Reserved
ISBN #978-1-935700-53-1
www.PeriscopeFilm.com

www.ingramcontent.com/pod-product-compliance
Lightning Source LLC
Chambersburg PA
CBHW080052190426
43201CB00035B/2184